T0302219

Mindful Safety

Mindful Safety

A Multi-level Approach to Improving Safety Culture and Performance

Christopher Langer

CRC Press is an imprint of the
Taylor & Francis Group, an **informa** business

First edition published 2021
by CRC Press
6000 Broken Sound Parkway NW, Suite 300, Boca Raton, FL 33487-2742

and by CRC Press
2 Park Square, Milton Park, Abingdon, Oxon, OX14 4RN

ISBN: 978-0-367-56502-2 (hbk)
ISBN: 978-0-367-56505-3 (pbk)
ISBN: 978-1-003-09805-8 (ebk)

Typeset in Times
by codeMantra

This book is dedicated to all those who consider themselves part of the 'elf and safety' brigade, but believe the profession has far more to offer than is often portrayed.

Here's to all the hard work you do to persuade your lords and masters that good health and safety practices prevent accidents and drive performance.

Remember that there is no business without health and safety, there is no safety without health and there is no health without mental health.

Contents

Foreword

My experience with addressing complex problems in organisations suffering from fatalities, serious injuries and high incident rates involving levels of safety, has led me to recognise the importance of focusing on the quality of human relationships. Quality means the degree to which employees feel included and a sense of belonging. Employees frequently offer the information we need to prevent serious accidents or fatalities, but those in authority are unable to hear it or see it.

Chris Langer's book helps us to understand the state of mind that must be achieved for everyone in the organisation to work towards listening with an open mind and maintaining the necessary awareness of their environment to spot both the human and technological signs of stress. To achieve this objective, he makes a strong case for integrating mindfulness in our safety and health toolkit through a combination of applied research, real-world examples and specific exercises.

Mindfulness has traditionally been an individual activity. Langer asks us to envision it in a much larger scope, using his M4 framework. This takes a holistic approach by looking at mindfulness on four levels: individual, relational, organisational and societal. I found the term 'relational mindfulness' especially useful, because the quality of relationships, and how people interact with each other, are seen as holding the potential to prevent accidents. His aim is to utilise relationships at work to enhance safety awareness and enable better responses to the potential dangers.

He states:

> This shifts the emphasis from attempting to change people's behaviour according to the prevailing interpretation of what is 'safe' or 'unsafe', to supporting activities that promote risk awareness.

Using M4 to analyse significant accidents, such as Air Florida Flight 90, Deepwater Horizon and Fukushima, provides new insights. Rather than placing blame, Langer looks at how increased mindfulness at each level can be applied at the

- Individual level – how can I keep myself present and aware?
- Relational level – whom are my actions affecting?
- Organisational level – which departments need to be in communication?
- Societal level – what external pressures are placing results or comfort ahead of people's welfare?

From this perspective, the loss of situational awareness described in many accident investigations can be reduced with mindfulness training.

The Deepwater Horizon investigation traced back the overarching failure to management, but how exactly did the management fail? The commission's report spells out many reasons, but Langer delves into the complex issues that we struggle with every day. These are the same issues that will most likely show up in the next failure. How can we encourage people to speak up in order to prevent failures? How do we stay present during routine procedures so that we see what is real and not what we expect to see? How can we question our own assumptions and those of the people in authority?

Every leader intuitively knows that health and safety requires 100 per cent engagement. A stressed-out workforce, driven hard to meet production targets, is unlikely to provide a sound basis for a positive health and safety culture. Mindfulness provides the mental toolkit to respond effectively to those pressures, as Langer has demonstrated in his work with clients. He suggests that a mindfulness programme would have made people more aware of the mental state of the German pilot Andreas Lubitz, before he purposefully crashed his plane in 2015, killing 150 people.

Finally, Langer states that this is not a self-help book, but it can actually serve as one, because we all need reminding to keep working on our ability to focus and stay in the present. It includes many valuable and practical tips for self-care, such as getting enough sleep and digital detoxing. It is my belief that this book applies to all levels of the organisation. Leaders must begin by finding their own ability to be mindful and then support others in that endeavour.

Rosa Antonia Carrillo
Safety Leadership and High Team Performance Consultant

Rosa Antonia Carrillo, M.S.O.D., President of Carrillo and Associates, is an internationally recognised consultant, speaker and author on leadership development and building high-performance teams with a focus on safety, environment and health performance. Her work is widely read on the topics of safety leadership, culture and relationship-centred leadership.

Preface

An ounce of prevention is worth a pound of cure.

– Benjamin Franklin*

I was once lucky enough to visit San Francisco and delighted in riding its iconic cable cars. Forget 200 mph supercars, I never imagined that hanging onto the side of these anachronistic, lovingly maintained vehicles whilst reaching a grand top speed of just under 10 mph could be so exhilarating. Ding ding and you are off, climbing the steep hills at a refreshingly sedate pace, enjoying panoramic views of a city that has provided the exquisite backdrop to many movies.

As you stand upright on the platform, you are immediately conscious of the street moving past just a few inches beneath your feet. You must learn to hang on. Riding a cable car is not without its safety risks even at that stately speed – after all, the technology can be traced back to the 1890s. Naturally, there are no seatbelts or airbags in sight, and so it pays to take good care when you make a journey. If the car comes to a sudden stop, you may find yourself being catapulted forwards. And you certainly do not want to fall off, get badly bruised, break an ankle, or worse, end up with a severed limb – yes, that really happens!

In fact, the US Department of Transportation recognises cable cars as one of the most dangerous forms of mass transit in terms of accident proneness by vehicle mile. One wonders if passengers should receive safety briefings before they board. On average, there is an accident every month, and lawsuits have been known to run into several million dollars.[1] None of this is intended to warn you off ever travelling by cable car, but, as you enjoy this unique mode of transport, just be mindful of the risks. Forewarned is forearmed, so observing the dangers means you are far less likely to become an accident victim.

One point that this book consistently makes is that to avoid injury, we must adopt a 'preventative mindset'. This is essentially about spotting the risks far earlier than is normally the case and equipping people with the mental tools to closely attend to their environment. All too often, we wait for something horrible to happen before acting. If there is one lesson worth pressing home, it is that turning a blind eye is usually a costly mistake.

Take a moment to consider the origins of the aforementioned cable car. Unsurprisingly, the catalyst for introducing the 19th-century technology we still see in operation today was an accident. Andrew Smith Hallidie, a talented Scottish engineer, watched as a passenger-carrying horse car struggled up Jackson Street on a damp, windy San Francisco day in 1869.[2] One of the horses slipped on the wet cobblestones, creating a horrific spectacle as the other horses fell and were dragged backwards down the street. Entangled and unable to free themselves from their

* This quote apparently appeared in the 4 February 1735 edition of the Pennsylvania Gazette. Franklin was apparently writing anonymously as an 'old citizen'.

rigging, they suffered broken legs and had to be destroyed. Thankfully, the passengers were unharmed, but it could have been far worse.

Hallidie's friend, Joe Britton, who was accompanying him at the time, said: "Andrew, why don't you put that wire rope of yours to use pulling these cars and prevent these accidents?"[3]

Britton's statement implies such accidents were not uncommon, but it was Hallidie's persistent efforts to stop them happening in the future that changed San Francisco for the better. His conception, plan and promotion for the first known mass transportation scheme to navigate hills safely made him a wealthy man. Though far from being the safest mode of transport in modern-day terms, his intuitive, technological leap forward made life far safer for the city's inhabitants than its cable-free, horse-drawn predecessor.

It makes one wonder how many other accidents could be prevented through astute observation, a mindful appreciation of the risks and a determination to change things for the better. Certainly, that thinking is in keeping with the spirit of this book. Establishing a preventative mindset towards health, wellbeing and safety is a gargantuan task and one that intimately involves individuals, teams, organisations and the rest of society.

Ever since the Industrial Revolution, the introduction of new technology has been a double-edged sword, often replacing human labour and driving up productivity, whilst creating the potential for accidents that can injure – or kill – on an industrial scale. Great strides have been made in reducing the number of accidents in all walks of life, but we are not out of the woods yet. If any argument were needed against complacency, the International Labour Organisation reports that there are 2.78 million fatal accidents at work each year and 374 million non-fatal work-related injuries and illnesses that result in extended absences from work.[4]

This book is dedicated to the application of mindfulness in the field of health, wellbeing and safety. I hope you will join me in aiming high with fresh, innovative thinking and clear goals in mind.

In pursuit of mindful safety, it is not hyperbole to suggest that:

- The number of fatalities could be reduced by a million worldwide.
- Non-fatal, work-related injuries and illnesses could be reduced by 100 million.

To achieve these goals, we won't be able to rely on the recycling of old thinking, such as behaviour-based safety initiatives from the 1980s. Fundamental change is long overdue to break through the performance plateaux.

Welcome to a promising new frontier, where the teaching of self-awareness, focus, emotional resilience and productivity comes first. And welcome to a practical approach that breaks firmly with the past, putting health and wellbeing at the heart of efforts to guard against workplace injuries and accidents. With a vision of optimal safety, the future is brighter, healthier, safer and smarter. Far from being an abstract vision, it is one that you can actively participate in – it embodies new thinking and new ways of doing things too.

Author

Chris Langer consults with business leaders and safety professionals to deliver cutting-edge training to teach improved mental focus and safety performance. He has worked with a diverse range of international safety organisations in the rail, road, aviation and construction sectors for 15 years. In this time, he has effectively utilised his skills in the field of human factors and occupational psychology, and he is a frequent contributor to the media, advising TV production companies on safety issues. He is also the founder of the M4 Initiative.

Chris began to realise that contemporary safety approaches were missing something critical. After becoming a mindfulness trainer, he saw an untapped opportunity in the synthesis of safety thinking and mindfulness. By teaching concentration through specific exercises to enhance resilience, safety incidents could be reduced substantially, saving staff from harm and eliminating unwelcome costs.

Proactively enhancing mental focus, whilst reducing safety risk and psychological stress is the mission of the M4 Initiative, which is set up to address these issues in the most practical way possible. Mental habits and behaviours are much more likely to be changed for the better if proven mindfulness techniques are used.

Outside work, Chris lives in the market town of Hitchin and can often be found with a tennis racket in his hand, or behind a camera lens. At home, he enjoys the rolling hills of Hertfordshire, but he has recently been exploring the incredible diversity of California. He also lived in Finland for six years. Though he was fairly normal before he left the UK, some have commented that a combination of the sauna and the long winter nights have rather messed with his brain.

Introduction

If you are reading these words and paying attention to them, you are practising a form of mindfulness. In a general sense, mindfulness can be defined as follows:

> The awareness that emerges through paying attention on purpose, in the present moment, and non-judgementally to the unfolding of experience moment by moment.[5]
>
> **(Kabat-Zinn)**

Most people would agree that paying attention to the task at hand, whether that is driving a car, piloting an aircraft, operating machinery or administering a drug to a patient, is pretty critical. And yet specifically training the mind how to pay close attention in such safety critical situations is incredibly rare. You can probably remember learning to drive and the instructor explaining how to drive safely, but they may not have wasted much time explaining how to prevent your mind from wandering. Instead, they probably just lunged for the steering wheel when you were about to hit a concrete bollard!

We spend much of our lives on autopilot, barely remembering how we managed to arrive at our destination or how we got something done. Thinking more broadly, many work-related scenarios rely on some form of cooperation between individuals, teams and organisations – for example, the pilot needs his co-pilot to monitor the aircraft's instruments, and the doctor requires information from other medical professionals to make the best possible decisions. It is no exaggeration to say that lives depend on people collectively paying attention to what they are doing.

This book essentially defines a radical new approach to the field of health, wellbeing and safety thinking – the M4 Initiative. What does M4 actually mean?

Essentially, we are interested in the impact of mindfulness at four distinct levels:

- Individual (about you)
- Relational (about relationships)
- Organisational (about the workplace)
- Societal (about society).

Though distinct, these levels are naturally interconnected, because none of us truly live in a bubble – even a teenager in headphones crossing the road, oblivious to the oncoming traffic. If we can simultaneously achieve mindfulness at all levels, we are well on our way to our goal of achieving a mindful safety culture. This is a radical idea because the practice of mindfulness has never before been systemically applied to safety critical business environments. Neither has it tended to embrace the world beyond our own heads – we will be redressing the balance here.

Our minds do not exist in a vacuum, because what we think and say, and how we behave, clearly affects those around us, and vice versa. Adopting a multi-level approach may seem a little daunting at first, but you'll quickly get the hang of it. It is a bit like viewing the world through different coloured lenses, which we can freely pick up to expand our perspective. The M4 Initiative rewrites the rules to embed fresh thinking in organisational culture, bringing with it the potential for much greater health and wellbeing. In these pages, you will find out how it can harness the power of mindfulness to improve focus and performance, whilst avoiding the harm and cost of safety incidents. Beginning with the efforts of individuals, it has the power to radically change teams, organisations, communities and even societies.

BEYOND BINARY THINKING

A growing body of scientific research supports the use of mindfulness for improved physical and mental health, as will become apparent. Evidence also suggests that it may reduce the number of safety incidents beginning to appear. Embracing holistic thinking is a necessity, because it is a huge mistake to split health and safety into separate compartments.

Nevertheless, many professionals in the field would still prefer to divide the world up this way. Rather than assuming the two are a rather odd couple, we need to treat them as being inextricably connected. Without the benefit of a more inclusive approach, we will produce stale interventions that fail to make a real difference. Safety management without reference to health and wellbeing is nonsensical, albeit all too common.

Binary health and safety thinking short changes organisations that possess a sincere desire to improve the lives of their employees. It will become abundantly clear that improving the health and wellbeing of workers also reduces the potential for safety incidents. How can someone act safely if they are not mentally or physically healthy? Just notice, for instance, how a mentally agitated driver will take greater risks on the road: speeding, running red lights or cutting other drivers up. We would naturally call the driver unsafe, but our knowledge of his mental state can help us dive deeper for explanations.

We can talk about the unsafe behaviours that may have contributed to accident statistics, but mindfulness takes us right back to the importance of mental processes. What goes on in people's heads is what ultimately determines their safe – or unsafe – behaviour. A calmer driver tends to be a better driver and is typically less prone to accidents. Behaviour is considered secondary in this sense – an outcome rather than a root cause. More to the point, calmness is a state of mind that can be taught to create the conditions for safer driving.

A Unified Approach

Achieving a higher level of mindfulness amongst individuals is a very worthy aspiration, but we are also obliged to pay attention to relational, organisational and societal factors. The M4 approach delivers change by training individuals to have a positive impact on those around them, as calmer managers create less stress in the workplace. As Mahatma Gandhi said, "You must be the change you wish to see in the world." In a slightly broader context, society and organisations can get inside our heads and modify our thinking through their values, rules and norms. By spotting how they do this, we can ask ourselves whether we accept the prevalent thinking and status quo, with a view to changing it if there is good reason to.

I would encourage anyone with an interest in the health and wellbeing of the people around them to apply M4 thinking. Health and safety practitioners, general managers, engineers, HR managers, directors and company owners are all responsible for ensuring people stay safe, remain in good health and perform optimally. Whatever role you play, the next section is designed to provide you with an outline for each level of mindfulness and how it relates to the field.

THE FOUR LEVELS OF MINDFULNESS

Level One: Individual Mindfulness

Here we are talking about the moment-to-moment awareness experienced by the individual. The key principles taught through formal practice are: staying anchored in our breathing, being aware of our bodily sensations and being aware of what our mind is doing. Direct experience and the suspension of our normally judgemental thinking are emphasised in this approach. An everyday awareness is fostered for superficially mundane activities, such as drinking a cup of tea, washing the dishes or walking to the shops. Most adults will also carry out potentially hazardous activities, such as driving, and this is where paying attention mindfully becomes more critical. Exposed to the risk of an accident each time they sit behind a steering wheel, being fully present is the key to remaining safe and out of harm's way.

Level Two: Relational Mindfulness

For there to be any meaningful communication at all, we need to listen and attend to other people. A growing body of evidence highlights how important relationships are for safe working conditions. It has been known for a long time that the relationship with one's manager can be a source of stress, but less attention has been paid to the effects poor relationships may have on safety. Addressing this deficiency, we will be emphasising how healthy relationships can positively affect the bottom line and reduce the number of safety incidents. This is the natural outcome of better communication, less interpersonal conflict and greater understanding.

Level Three: Organisational Mindfulness

It is argued here that organisations will need to demonstrate mindful leadership in order to improve the health and wellbeing of their employees. Stretching beyond a narrow focus on production disruptions and non-compliance costs makes good business sense. Frontline workers often know more than their managers about the reality at the sharp end, so their knowledge and expertise should be sought. They can be harnessed as active participants in reducing the number of safety incidents, increasing efficiency and improving the organisational culture.

There probably isn't much to be gained from a hair-splitting debate on the differences between 'organisational culture' and 'safety culture', since it is likely that the same values, attitudes and beliefs underpin both. A somewhat outdated view of things tends to treat 'safety culture' as a separate entity, but it makes more sense to embed processes that integrate business and safety. After all, inherently unsafe acts are unlikely to be good for business. If managers and frontline workers share similar perceptions of the prevailing organisational culture, it implies a higher degree of organisational mindfulness.

Level Four: Societal Mindfulness

The influences of national and local culture, and technology, can all play prominent roles in this domain. The cultural traits of the Japanese had a significant part to play in the nuclear meltdown at Fukushima (see Chapter 8). And we all love smartphone technology these days, but attention lapses can have disastrous consequences for road users. Being mindful of how these larger 'forces' affect our thinking is critical if we are to avoid the accidents of the future. Making the best possible decisions in safety critical environments may sometimes mean opposing the societal norms that embed themselves in our thinking. In a similar vein, our health may be affected if we cannot use the technology at our disposal wisely or effectively ration its use.

WORKING WITH THE MULTI-LEVEL APPROACH

In practice, there will usually be circumstances where one level features more prominently than others, as highlighted by the text and commentary.

For example, if we are talking about failing to concentrate whilst driving, the most appropriate level would be the individual one. But if we are talking about a dysfunctional relationship between a supervisor and employee, the relational level of analysis would be better suited.

In more complex situations, such as in the event of a safety incident, each level can also be used as a kind of lens to generate understanding and perspective. This would be appropriate where multiple factors on several different levels are implicated. As well as providing perspective, this can usefully counterbalance the view that human error is largely to blame when things go wrong. Many incident investigations prefer to focus almost exclusively on the individual level, but this is counterproductive if systemic factors are effectively ignored. In this book, you will notice that each chapter ends with a section on how to apply the learning at each of the four levels.

A NOTE FROM THE AUTHOR

You have to be slightly geeky to write a book like this. As a student, long before electronic searches were possible, I can remember spending hours chasing down journal articles in the library, always searching for the best thinking. I hope I have brought some of that ardour to these pages. As a self-confessed student of psychology for more than 25 years now, my desire to apply the best of psychology in the real world remains undimmed.

Whilst doing the research for this book, I read through many thousands of pages of accident reports. Some of it, I have to tell you, was very grim! One could label me as having a rather morbid interest in plane crashes, train derailments, oil rig explosions and nuclear meltdowns. However, there is always good reason to hope for better outcomes in the future, even in the midst of unimaginable human tragedy. As Voltaire put it, “Every evil begets some good.” This is only true if we can learn the lessons to avoid the same mistakes recurring. If this book contributes in some small way to fewer accidents, it will have served its purpose.

My interest in mindfulness is a relatively recent phenomenon. Originally a sceptic, I persisted with the practice, slowly becoming convinced of its transformative effects through personal experience. A few years ago, I began thinking about how it could make a real difference in safety critical environments, where very little applied research had been done. In fact, my main motivation for writing this book stemmed from a desire to fill the void in the resources currently available. Writing in the absence of a greater body of research is a necessary evil sometimes.

As long as our minds need to concentrate, mindfulness will always be important. If yours begins to wander during the reading of this book, remember to bring it back to the task at hand!

HOW TO USE THIS BOOK

This book is designed to be read from start to finish, but you may wish to go directly to the parts that are of most interest to you. The outline below tells you what you can expect to find in each chapter. Case studies are used throughout to highlight the risks, alongside practical tips and advice for avoiding incidents.

Chapter 1 – Mindful Safety Culture. We are living in the age of global pandemics, and the world has changed, ushering in new practices in health and safety. This chapter takes a look at conventional definitions and approaches to safety culture and explains the impetus for change. It makes sense to aspire to a new kind of safety culture: mindful safety culture. Not only does this reinvigorate the debate, but it also describes what we are aiming for if we want to simultaneously improve both safety culture and safety performance. The key attributes of mindful safety culture are described here.

Chapter 2 – Safety in Four Dimensions. Readers wishing to grasp the principles and thinking behind the M4 Initiative will find this chapter indispensable. The four levels and their relevance to health, wellbeing and safety are explained, providing a clear framework for the rest of the book. After reading this chapter, you won’t have any difficulty explaining to someone else what the M4 approach is all about, and why it is fundamentally different from every other approach out there.

Chapter 3 – New Tools for Incident Investigation shows how the M4 approach can be comprehensively applied to investigate safety incidents. Using the case study of Air Florida 90, new insights can be gained by examining the accident from the four different levels. Adopting a multi-level perspective is important if we are to maintain a fair, objective assessment of any major safety incident.

Chapter 4 – Self-Care: The Cornerstone of Mindful Safety is all about how we can look after ourselves better. It describes how health can be significantly boosted with reference to the following topics: burnout, sleep, exercise, digitally detoxing and ‘being present’. If you apply just some of the tips in this chapter, you’ll start to notice a big difference, both at home and work.

Chapter 5 – Fatigue: Safety's Silent Saboteur examines the detrimental role of fatigue in leadership and performance. How fatigue can affect safety critical decisions is discussed with reference to what can go wrong in safety critical environments. This chapter is packed with plenty of practical tips on what you can do to counteract the effects.

Chapter 6 – Distracted Minds, Lost Lives is an important chapter because it is the area where mindfulness can assist the most. There are myriad distractions in everyday life, and we need to train our minds to stay on track, especially if we are responsible for the safety of others and there is the risk of an accident. Coming off autopilot and maintaining situational awareness are essential.

Chapter 7 – The Mental Health Elephant at Work tackles a challenging subject with the aim of reducing the associated stigma. Mindfulness is a highly effective way of preventing mental ill health, and the research evidence highlights the benefits. If you want to know how to prevent your stress levels from rising, and help others under pressure, this is the chapter to read.

Chapter 8 – Culturally Mindless: The Ostrich Syndrome explains why we absolutely mustn't bury our cultural heads in the sand when it comes to safety. By becoming aware of societal and cultural norms, we can make safer decisions and may even be able to prevent a nuclear accident. The subject of how to 'unfollow' the cultural herd in order to reclaim your ability to think independently is also tackled here.

Chapter 9 – Speaking Up to Avoid Catastrophe demonstrates why reporting concerns is so important, and why those in authority need to listen. If you want to design a reporting system that really works to capture important concerns that otherwise might be missed, then look no further. This chapter takes a look at how to overcome the barriers to reporting, as well as the role of confidential and anonymous reporting to ensure valuable safety intelligence is allowed to surface.

Chapter 10 – Mindfully Learning from Positives demonstrates how the brain's negativity bias can be overcome for much greater positivity. The potential to learn from positives cannot be underestimated – it very often represents a huge, organisational learning opportunity. Better safety outcomes are achievable by fostering good relationships and positive safety scripts, thus highlighting a new direction of travel for safety thinking.

Chapter 11 – From Blame to Safety Enlightenment describes exactly what we can do to avoid the blame game and create a working environment that promotes effective collaboration. The goal is an enlightened workplace culture to provide the right basis for a much healthier, safer environment. Implement the suggestions in this chapter and they could fundamentally change the nature of your relationships and workplace.

Chapter 12 – Mindfulness Training for Improved Safety Performance provides all the examples you need to understand the well-researched impact of mindfulness-based safety programmes. Greater alertness, improved focus and enhanced performance are just some of the benefits. Case studies with the nuclear power industry, London bus drivers and other sectors illustrate how this can be achieved in practice.

NOTES

1. San Francisco cable car accidents cost city millions. Fox News. 14 April 2013. Available from http://www.foxnews.com/us/2013/04/14/san-francisco-cable-car-accidents-costs-millions.html (Accessed 22.07.2018).
2. Jansen, J. (1995). San Francisco's Cable Cars: Riding the Rope through Past and Present. San Francisco, CA: Woodford Press.
3. Ibid.
4. Safety and Health at Work. Available from http://www.ilo.org/global/topics/safety-and-health-at-work/lang--en/index.htm (Accessed 21.07.2018).
5. Kabat-Zinn, J. (2003). Mindfulness-based interventions in context: past, present, and future. Clinical Psychology: Science and Practice, Volume 10, pp. 144–156.

1 Mindful Safety Culture

The term 'safety culture' first gained prominence in 1984 after the Bhopal disaster, and became increasingly popular after the world's worst nuclear accident at Chernobyl in 1986. But a few decades later, a radical overhaul of its existing assumptions is needed if safety culture is to remain relevant. We will need to take stock of the various definitions and approaches and move forward, especially in the light of the Covid-19 pandemic.

A solid reference point is the idea that safety culture expresses 'the way we do things around here in relation to safety'. Amongst both practitioners and academics, there is a consensus that safety culture reflects a proactive stance to improving safety in operational environments. It is also instructive to consider an accident where there is widespread agreement about the lack of safety culture as a key underlying cause – one such example is Deepwater Horizon.

Case Study: Deepwater Horizon and the Absence of Safety Culture

Eleven lives were lost; five million barrels of oil were spilled into the sea; BP spent tens of billions in fines, alongside economic claims, disaster response efforts, and clean-up and restoration programmes, not to mention the massive reputational damage inflicted.

President Barack Obama created the National Commission in 2010, shortly after the disaster, for the purpose of independently and impartially investigating the causes of the oil spill in the Gulf of Mexico. The troubling symptoms of a disaster-in-the-making could have been picked up far earlier with a much stronger safety culture. The National Commission drew attention to the industry's safety culture in its report:

> The immediate causes of the Macondo well blowout can be traced to a series of identifiable mistakes made by BP, Halliburton, and Transocean that reveal such systematic failures in risk management that they place in doubt the safety culture of the entire industry.[1]

How did the crew at Macondo come to describe it as 'the well from hell', whilst BP's Vice President of Drilling Operations said it was "...the best performing rig that we had in our fleet and in the Gulf of Mexico?"[2] You could be forgiven for thinking that they were talking about different oil rigs altogether. The oil rig crew and senior management were not only talking a different language, but their views on safety culture appeared to be at opposite ends of the spectrum.

What we know from the National Commission's report is that key safety systems were intentionally switched off. For starters, the physical alarm system on the rig was disabled a year before the disaster. A crucial safety device to shut down the drill shack if dangerous gas levels were detected was also disabled, or 'bypassed'. This last fact had not gone unnoticed – indeed, the Chief Technician had previously protested to his Supervisor. The response he received was truly astonishing, indicating a much wider malaise: "Damn thing been in bypass for five years. Matter of fact, the entire fleet runs them in bypass."[3]

The catalogue of management failings did not end there. There was no procedure for running, or interpreting, what in the oil and gas business is called the 'negative pressure test' to show that the well was safely sealed with cement. The crew were not, therefore, able to decipher critical data that would have alerted them to the danger signs. To make matters worse, there was no procedure for calling back to shore for a second opinion about confusing data. And there was no formal training for the crew, especially in response to emergency situations.

By far the biggest failing was the failure to learn from a near-miss incident in the North Sea just four months earlier. The basic facts of the two incidents were essentially the same, but the North Sea near miss did not reach the level of catastrophic blowout. Tragically, the lessons from the North Sea incident weren't communicated to the crew at Deepwater Horizon. The critical learning remained frustratingly 'locked away' in the system. Had this learning reached the right personnel in time, it may have prevented the disaster.

An 'operations advisory' containing the critical information was sent to some of the fleet in the North Sea, and a PowerPoint presentation was created for the purposes of learning from the incident. Neither made it to the Deepwater Horizon crew. This fitted part of a pattern of "...missed warning signals, failure to share information, and general lack of appreciation for the risks involved".[4] In these circumstances, the very notion of safety culture seems quite alien.

SAFETY CULTURE: CONTENT AND CHARACTERISTICS

The field draws from a range of contributions, often using 'culture' in a more general sense as a departure point. Several notable writers have added their own unique perspectives – for example:

- Hofstede (1991) speaks of culture as the collective programming of the mind, a kind of 'mental software' that distinguishes one group of people from another.[5]
- Bang (1995) suggests organisational culture is a set of common norms, values and world views that emerge when an organisation's members interact with each other.[6]
- Reason (1998) analyses safety culture in terms of five interlinked subcultures (informed, learning, reporting, just and flexible cultures) based on incident analyses.[7]
- Guldenmund's (2000) interpretive model contains three layers: unconscious and unspecified basic assumptions, espoused beliefs and values, and artefacts.[8]

A recent review of the safety culture literature by Cooper (2016) over the previous 30 years has usefully described an emerging consensus from academic research and public enquiries into safety disasters.[9] In Cooper's view, there are six major safety culture characteristics:

1. Management and supervision (e.g. visible safety leadership)
2. Safety systems (e.g. formalised strategic planning)
3. Risk (e.g. risk appraisal, assessment and controls)
4. Work pressure (e.g. safety versus productivity)
5. Competence (e.g. knowledge, skills and ability of people)
6. Procedures and rules (e.g. codified behavioural guidelines).

These six characteristics are no doubt a useful starting point in highlighting some of the fundamentals of safety culture. But far less is said about the agents of safety culture change, or the psychological tools needed to change or improve any of the characteristics.

As a result, there is a real need to think beyond analytical approaches that pay scant attention to the psychological flexibility required to make positive changes. Here, we are shifting the emphasis to the psychology of habit formation, and what needs to happen through awareness to change unsafe behaviours into safe ones.

Mindfulness practice, which creates the awareness required to change, can be used as an effective 'habit releaser'. It seems pointless to talk about change or improvement in safety culture, without providing the tools or agency for it to occur in the first place.

SAFETY CULTURE IN THE AGE OF PANDEMICS

The future of safety culture is incredibly important if we wish to meet new challenges, especially in the era of global pandemics. It is an ideal time to scrutinise the whole concept and improve upon it.

The M4 approach takes the view that safety culture is much more than just shared assumptions, norms, values or a set of characteristics. With the right tools and a multi-level approach, it is possible to change and improve safety culture. A great many current approaches focus on content, rather than process, and rarely discuss the agents of change. Without a doubt, this is a missed opportunity.

We must be able to understand the process of how people come to think and act the way they do, with the goal of fully enabling change. The psychological tools for achieving this are provided by the mindfulness approach. Deploying these tools helps to 'unfreeze' existing safety cultures, whilst providing the catalyst for new action. We are interested not only in the safety of workers, industrial processes and procedures, but also in the safety of whole societies.

To set the bar higher, we need safety leaders to be not just visible, but mindful. We need risk awareness to be present in an everyday sense, just as much as we need risk assessments and controls. We need safety rules and procedures to be mindfully enacted.

And when they need changing, this needs to involve a high degree of consciousness to break old habits and form new ones.

The Covid-19 pandemic has clearly demonstrated how safety is of everyday relevance to every worker and citizen globally. Commitment to safe behaviour is required by everyone, whether they are working or not, and a unified 24/7 effort is necessary. To reduce the rates of transmission and infection, national governments, organisations, workers and citizens must work with the same safety goals in mind.

The pandemic's hammer blow to the world on multiple fronts fundamentally affected our everyday existences and restricted our freedoms. It reached indiscriminately into every home, office and business across the globe. Workplace health and safety has undergone a big transformation as a result, and views of safety culture will need to adapt accordingly.

MINDFUL SAFETY CULTURE

The world changed significantly in 2020, and our ideas about safety culture must follow suit. It makes sense to shift the emphasis to what it takes to create and sustain a 'mindful safety culture'. Mindful safety culture can be defined as follows:

> the degree to which an organisation's people – individually, relationally, organisationally, and societally – consciously direct their everyday attention to improving safety.

In this approach, assumptions, values and beliefs do not represent the destination, but are viewed as reference points along the route to an improved safety culture. It is important to acknowledge their influence and bring them into conscious awareness wherever possible. There will of course be a shared understanding of how 'things are done around here' in relation to safety. This is important for establishing a suitable frame of reference. But ultimately, we are more interested in the process by which a given safety culture has come into existence, because this can provide the clues to its future, and crucially, its ability to change.

A mindful safety culture has reflective, mirror-like properties to enable the noticing and observing of oneself and one's organisation. Through an honest appraisal of what works and what doesn't work, it is continuously developing with both safety and performance in mind. This process is willingly owned just as much by frontline workers as it is by senior and middle management. The dialogue between these groups is essential to improve safety culture.

The impact of relational factors on safety is a hugely neglected area. Engagement is usually achieved through productive dialogue. Research by Lockwood (2007) suggests that engaged employees are five times less likely to experience a safety incident, and seven times less likely to have a lost-time safety incident than non-engaged employees.[10] This indicates that the teaching of so-called 'soft skills' to promote mindful relating is likely to have hard results which affect the bottom line.

EIGHT KEY ATTRIBUTES OF A MINDFUL SAFETY CULTURE

1. **Focuses on the change process itself**. A lot of safety culture thinking is focused rather idealistically on the 'right' values and behaviours, but not on the process through which these are determined in the first place. To mindfully change habits and behaviours, we need to understand how they are formed, and then how to release ourselves from their grip.
2. **Conscious cultural reprogramming**. Attitudes, habits and behaviours are formed in any safety or organisational culture. By identifying where autopilot occurs in safety tasks and procedures, we are essentially becoming mindful. A conscious reprogramming of attitudes, habits and behaviours is far more likely to ensure successful change and improvement in safety culture.
3. **Embodies 'felt-change' and continuous development**. There is nothing new in the idea of continuous organisational change and development. But to change a safety culture, it would be advantageous to focus far more on cognitive and inner processes. To achieve a real change, we must be cognisant of what is required for people to experience change in their own minds.
4. **Provides psychological tools for resilience and flexibility**. To effect a change in safety culture, we need to focus our efforts on how we can overcome adversity and challenging situations. It involves much more than extolling the virtues of the latest management fad. Equipping an organisation's people with the right psychological tools takes resilience and flexibility from being mere soundbites to operational realities.
5. **Embraces the four levels**. Becoming mindful can happen in our own heads, in dialogue with others, or at the organisational and societal levels too. Thinking four dimensionally in terms of the individual, relational, organisational and societal is a prerequisite for safety culture change. The Covid-19 pandemic provides ample proof that to effectively change safety behaviours, we need to embrace all four levels.
6. **Unifies safety culture and safety performance**. There is no special conflict implied between safety culture and safety performance, because paying attention to cognitive and emotional processes improves performance and doesn't need to come at the expense of safety. Think, for example, of a pilot who is paying close attention to his plane's instruments and the working environment, whilst monitoring his emotional state. He or she can maintain a high state of alertness and perform well at the same time, whilst calmly communicating with other professionals.
7. **Embeds good mental health**. If frontline operatives and senior managers are not mindful of the importance of mental health, how can we reasonably expect people to perform their jobs safely? Artificially separating mental health and safety culture can have tragic consequences, if distressed or traumatised individuals work unsafely. Positive mental health for all an organisation's people is seen as a realistic and achievable goal.

8. **Works constructively with ambiguity**. Any process that provides a catalyst for change is likely to evoke a sense of ambiguity. As old habits and ways of working are challenged, and new attitudes, behaviours and habits formed, there is a need to tolerate ambiguity whilst things are in a state of flux. Many people find change tough psychologically, but the skills needed to work with ambiguity can be taught to ensure that safety culture improvements are more easily embedded.

NOTES

1. National Commission on the BP Deepwater Horizon Oil Spill and Offshore Drilling (2011). *Deep Water: The Gulf Oil Disaster and the Future of Offshore Drilling*, p vii. Available from https://www.gpo.gov/fdsys/pkg/GPO-OILCOMMISSION/pdf/GPO-OILCOMMISSION.pdf (Accessed 15.06.2020).
2. Ibid., p. 6.
3. Pilkington, E. (2010). Deepwater horizon alarms were switched off 'to help workers sleep'. *The Guardian*. 23 July 2010. Available from https://www.theguardian.com/environment/2010/jul/23/deepwater-horizon-oil-rig-alarms (Accessed 15.06.2020).
4. National Commission on the BP Deepwater Horizon Oil Spill and Offshore Drilling (2011). *Deep Water: The Gulf Oil Disaster and the Future of Offshore Drilling*, p. ix. Available from https://www.gpo.gov/fdsys/pkg/GPO-OILCOMMISSION/pdf/GPO-OILCOMMISSION.pdf (Accessed 15.06.2020).
5. Hofstede, G.R. (1991). *Cultures and Organisations: Software of the Mind*. London: McGraw-Hill.
6. Bang, H. 1995. *Organisasjonskultur (3.utgave) [Organizational Culture]*. Oslo: Tano AS.
7. Reason, J. (1998). Achieving a safe culture: theory and practice. *Work & Stress*, Volume 12(3), pp. 293–306.
8. Guldenmund, F.W. (2000). The nature of safety culture: a review of theory and research. *Safety Science*, Volume 34, pp. 215–257.
9. Cooper, M.D. (2016). *Navigating the Safety Culture Construct: A Review of the Evidence*. Franklin, IN: BSMS.
10. Lockwood, N.R. (2007). Leveraging employee engagement for competitive advantage: HR's strategic role. *HR Magazine*, Volume 52(3), pp. 1–11.

2 Safety in Four Dimensions

Mindfulness means being awake. It means knowing what you are doing.

– Jon Kabat-Zinn[1]

Knowing what we are doing, both individually and collectively, at home and in the workplace, forms the unique basis of the multi-level M4 approach. In the introduction, each of the four levels was sketched out, but these levels are explained in much greater detail here, with a clear rationale for the thinking behind them. If you plan to read just one chapter, I suggest you read this one. Not only does it form the backbone of the rest of the book, but it also explains why mindfulness is so important for health, wellbeing and safety. What this chapter also explains is how the M4 approach differentiates itself from other approaches in the field.

INDIVIDUAL MINDFULNESS IN ACTION

Every safety critical task carried out by a human being requires awareness, concentration and attention. Whether we are talking about driving a car, bus, train or forklift truck, piloting an aircraft, operating on someone's heart or monitoring the control panel in a nuclear power station, mindfulness is a prerequisite for safe performance. If awareness is not brought to the task at hand, there is the ever-present risk of an accident: a collision, crash, failed operation or even a nuclear meltdown.

Take road accidents in the United States, for example, where there were over 40,000 fatalities in 2016 alone.[2] Mobile phone use has been cited as the number one driver distraction when it comes to crashes,[3] but there are many other distractions as well, such as talking to passengers, eating, or adjusting the radio or climate controls. Driver distraction may also involve cognitive factors such as attention, situational awareness, mental workload and risk perception. The case studies on the nuclear power industry and London bus drivers presented in Chapter 12 show how mindfulness training can achieve direct improvements in these areas. Such factors also play a role in the many other environments referred to above – for example, in healthcare, transportation and energy. In fact, it is difficult to see how 'paying attention on purpose' can be ignored for long in any safety critical environment.

Health is so often treated as the poor relation of safety. Much of safety literature devotes very little space to health, evoking a rather strange universe where people do not suffer with physical ailments or ever get depressed. The challenge is to raise the profile of health and wellbeing in safety critical environments, acknowledging that good health in a holistic sense will always underpin safe operations. The case of Germanwings 9525 (see Chapter 7), where a suicidal pilot locked himself in the cockpit and then crashed his aircraft into the Alps, is a tragic reminder that we especially cannot afford to neglect mental health.

Mindfulness teaches us to observe our thinking and moods, and it can therefore be used as a tool to notice low moods and prevent a slide into depression. In fact, the clinical evidence suggests it is an effective form of treatment for both anxiety and depression. Looking after ourselves is reliant on equal attention being paid to both the mental and physical aspects of our health, with the mind and body treated as an interconnected system. Chapter 4 takes a closer look at how we can best look after ourselves, and the potential consequences if we do not.

Most mindfulness training focuses on the individual, and this is a logical place to start. But unless we are performing a safety critical task in complete isolation, mindfulness needs to be examined in broader contexts. Many social interactions between individuals, such as those taking place between an air traffic controller and a pilot, or between a train driver and a signaller, require mindful communication to ensure optimal system safety.

RELATIONAL MINDFULNESS IN ACTION

> Safety is sustained as a priority within the context of relationships, making the creation of high-trust working environments among members and across the function of the organisation highly important.[4]

Relational mindfulness plays a critical role in ensuring the health and safety of the workforce. The evidence suggests that healthy relationships and high-quality social interactions between individuals can make a huge difference. In healthcare, for example, there is extensive evidence highlighting the correlation between the quality of the relationships amongst staff and the quality of the healthcare delivered.[5] Rows between surgeons can increase the mortality rate significantly, as they did in one cardiac unit at St George's Hospital in London.[6] A similar pattern of findings emerged in a case study of Southwest Airlines, where improved staff relationships led to better coordination and task integration.[7] Perhaps confounding the usual expectations, the more effective use of soft skills led to concrete, quantifiable results in that there were fewer flight departure delays. Who said soft skills were fluffy?

Open dialogue and participation are clearly important, too. In one study, a participative supervisory style not only improved social interactions, but it turned out to be the best predictor of work groups taking on safety initiatives.[8] It also correlated with fewer working days lost as a result of accidents. In order to foster open dialogue, a tolerance of multiple perspectives is required. Things may look very different on the frontline than they do from an office. Good listening skills and an openness to the experience of others are required. If a trusting atmosphere prevails, conversations about health and safety can shape people's understanding of reality, encouraging the appropriate actions. A mind closed to the perspective of others will end up thinking the same thing at the end of the conversation as at the beginning. What then is the point of the conversation? Closing one's mind means closing one's eyes too.

This is where the practice of mindfulness can positively shape such conversations. Listening to ourselves and any thoughts, feelings, sensations and intuitions helps us gain perspective. We can also make an authentic attempt to listen out for the same things in other people. This is called 'inquiry', and it applies equally to ourselves and to other people. In both cases, we are interested in the nature of experience, as no experience is inherently right or wrong – it just 'is'. We mustn't underestimate the challenge, because if a shared understanding of reality cannot be reached, it can spell disaster. The Deepwater Horizon oil rig blowout in the Gulf of Mexico claimed the lives of 11 workers in 2010 and also caused an ecological disaster. There was not much evidence of listening on the rig, open dialogue or tolerance of the crew's perspective from senior management. Some crew members had dubbed it the "well from hell".[9] BP's vice-president of drilling operations had, in stark contrast, called it the "best performing rig" in the fleet and in the Gulf of Mexico. The lack of a shared understanding of the safety risks on the oil rig showed up in the social interactions and the everyday language used by senior management and the oil rig workers.

ORGANISATIONAL MINDFULNESS IN ACTION

We will stay with the Deepwater Horizon disaster to illustrate what happens in the clear absence of organisational mindfulness. The National Commission's report provides the following commentary:

> Better management by BP, Halliburton and Transocean would almost certainly have prevented the blowout by improving the ability of [the] individuals involved to identify the risks they faced, and to properly evaluate, communicate and address them.[10]

In other words, management had a clear responsibility in the way events unfolded. Just a few weeks before the accident, a safety culture survey revealed that 46 per cent of crew members felt that the workforce feared reprisals if they reported unsafe situations.[11] Fifteen per cent felt that there were not always enough people available to carry out work safely. Safety had been sliding down the agenda at the expense of production targets.

Organisational mindfulness brings a quality of collective awareness to the prevailing safety culture, allowing space for objectivity and constructive criticism. It describes the conscious process of collectively perceiving and evaluating the operating environment's health and safety risks. Just as an individual can pay attention to what his or her mind 'sees' at any one time, an organisation can pay attention to what it 'sees' through the eyes of all its employees.

How does organisational mindfulness manifest itself? The following process characteristics determine the degree of mindfulness an organisation has attained.

Accepting Flux

Values, attitudes and behaviours can be held or enacted mindfully, or mindlessly. What matters more, perhaps, is how they can be moulded and adapted intentionally to the 'always in flux' reality of complex environments. This means going far beyond contemporary definitions of safety culture, which tend to emphasise fixed states more than the process of how to achieve them. Specifying a type of behaviour for a particular job is all very well, but what happens when the conditions require that behaviour to be modified? The goal here is to produce risk-aware individuals who choose to behave in the best interests of their own and others' health and safety, rather than automatons who blindly follow rules and procedures prescribed for a narrow range of circumstances. This shifts the emphasis from attempting to change people's behaviour according to the prevailing interpretation of what is 'safe' or 'unsafe', to supporting activities that promote risk awareness.

Learning from Positives Just as Much as Failures

Most organisations are preoccupied with failures, and forensic investigations often follow accidents to determine the root causes. This is not, of course, a bad thing as long as lessons are learned and applied in future situations. However, we can learn as much from positive events as from negative ones, if not more. There absolutely should be a strong desire to learn from lapses, errors and inconsistencies, but the pursuit of organisational learning from good practice, near misses, recoveries and mindful performance should be just as relentless. Organisations with global operations will need to rapidly learn to share such knowledge across national boundaries and time zones.

Listening to the Frontline

Frontline workers provide vital information on the current state of the system. Failing to listen, especially over prolonged periods of time, can cut an organisation off from the information it needs to operate without serious incident. Managers must show some sensitivity here and listen carefully to the experiences of their frontline workers. Fear of speaking up can indicate a blame culture. This subject is explored in much greater depth in Chapter 11.

Mobilising Resilience

An important feature of mindful organisations is their ability to mobilise resilience when unexpected events occur. Instead of being disabled by these, "knowledgeable people self-organise into ad hoc networks to provide expert problem solving".[12] It is not just situational awareness and the act of noticing that matters, but also the response under pressure that follows in extraordinary circumstances. Responding rather than reacting to such situations will help secure more successful outcomes. To bounce back from adverse or unexpected events requires the skillful regulation of one's emotions, something that can be taught.

Promoting Risk Awareness

Why do we need to promote risk awareness? There is a fundamental problem with relying solely on compliance with rules and procedures for the control of hazards. Hazard control strategies overemphasising compliance are likely to prove deficient, especially where the unexpected throws a major spanner in the works of standard operations. Instilling risk awareness into employees is therefore critical if the safest course of action is to be followed in any given situation. Accident investigations frequently find that employees did not know what operational rules to apply. This may be because the rules are fundamentally not fit for purpose, as highlighted by the Glenbrook train crash in Sydney, which claimed seven lives in 1999.[13]

In the inquiry that followed, one railway employee described the rules as "incredible waffle". Others said that they were "confusing, complex and overlapping". Every time an accident occurred, a new rule was promulgated with the intention of preventing the same thing happening again. However well-intentioned that was, it led to a behemoth of a rulebook, which ran into thousands of pages. Worse still, some employees felt the rulebook's primary purpose was the punishment of offenders for non-compliant acts. The weighty tome was certainly suggestive of being used to 'bash people over the head'.

A prime example of this verbosity was the procedure described for reversing a train. It originally ran to a whopping 12 pages, yet during the inquiry, one railway employee was able to reduce the procedure down to fewer than 40 words. Creating easy-to-follow, peer-reviewed procedures builds risk awareness based on shared understanding. Leaving such activities until after an accident has occurred is a lost opportunity to save lives.

SOCIETAL MINDFULNESS IN ACTION

If organisations can choose what health and safety risks they systematically focus their attention on, so too can whole societies. But in terms of influencing behaviour, the instruments of change are more likely to be government or regulatory bodies. During the Industrial Revolution, technological change created the possibility of multiple fatalities and serious injuries unheard of in an earlier age. Society was eventually forced to legislate for safer, healthier working conditions, thus instilling into the public consciousness the message that killing people in factories, cotton mills or coal mines was no longer acceptable. Neither citizens nor organisations exist in a vacuum, responding as they do to the broader attitudes, trends and patterns in society. Legislation is one of the ways society can get inside our heads and influence our thinking.

A societal regard for health and safety can maintain our alertness, instilling the risk awareness necessary to prevent accidents. Industrial accidents in most developed countries are far rarer than they once were, but we must remain alert to the early warnings signs, which are nearly always observable long before an accident. In the case of the nuclear meltdown at Fukushima (see Chapter 8), a greater degree of societal mindfulness could have overcome cultural 'blind spots' and addressed the safety risks years before disaster struck. This is always a formidable challenge when attitudes and behaviours are deeply embedded in the culture, but they can be changed. Road safety, public hygiene and alcohol abuse have all been successfully targeted in national campaigns.

In a different area entirely, societal attitudes to mental health have shifted substantially in the last few years. Mental health awareness is no longer a mere appendage to physical health (see Chapter 7). Many workplaces are openly embracing the challenge, recognising that stress, anxiety and depression can be alleviated through training to break down the wall of silence – mindfulness has a role to play here too.

OTHER THEORIES AND FRAMEWORKS

Over the last few decades, there have been several recognised approaches to health and safety, each with their own angle, theoretical principles and biases. We are interested in what these approaches bring to the table, whilst taking a look at their respective merits and pitfalls. Behavioural safety, human factors, organisational factors and Safety-II approaches are all placed under the spotlight. In practice, they shouldn't be seen as 'ideological strongholds', because some of the approaches overlap. Most health and safety practitioners will not treat them in isolation, even if the text unintentionally implies it is possible.

Also, whilst the M4 approach has largely been built from scratch, it will be clear that some elements owe their existence to the ideas of others. This is most noticeably the case in relation to Hollnagel's Safety-II approach, which emphasises learning from 'things that go right', just as much as from 'things that go wrong'. I've incorporated some of Hollnagel's original thinking into my approach, and I fully acknowledge his influence in this regard.

Behavioural Safety

The approach taken in behavioural safety management is largely derived from behaviourism, which dominated psychology in the first half of the 20th century. Though its basic assumptions came from an earlier time, behaviourism saw a resurgence towards the end of the century, in an effort to reduce the flatlining incident and injury rates. Rather than boiling down to a single method, behavioural safety management can be seen as a range of techniques for the improvement of safety performance. Key characteristics of this approach are: setting goals, measuring performance and providing feedback.[14]

At its heart is the assumption that a significant proportion of accidents are primarily caused by the behaviour of frontline staff. For example, pilots, drivers, production operators and maintenance technicians. It is sometimes claimed human error causes up to 80 per cent of incidents. Behavioural safety approaches tend to focus on reinforcing safe behaviours in the workplace, whilst seeking to eliminate the unsafe ones. Observation, intervention and feedback are all emphasised in an effort to modify behaviour.

Advocates of behavioural safety say that it can help identify dangerous situations and shape desired behaviours whilst demonstrating management commitment to improving workplace health and safety. If a trusting atmosphere prevails, it may also improve dialogue between managers and frontline staff. Critics argue that it completely neglects cognitive processes, with an 'operator focus' that excludes the impact of management decisions. It may largely ignore important organisational factors, such as inadequate training, poorly designed equipment, inappropriate rules or a lack of resources. The focus tends to be on readily identifiable behaviours, such as the wearing of personal protective equipment and the proper use of harnesses or ladders.

Though a balanced view of behavioural safety must acknowledge its usefulness under certain conditions, its inability to embrace the implications of low-probability, high-impact events is considered a flaw. Despite decades of behavioural safety initiatives, major accidents are still occurring. Its staunchest advocates seem unable to step outside its narrow premise, as doing so would expose the approach's inability to embrace the reality of complex, dynamic environments.

Nudge

Though largely informed by economic theory, and not specifically designed for the health and safety domain, Thaler and Sunstein's 'nudge' approach can be considered an extension of the behavioural approach.[15] Its application has been discussed by health and safety practitioners interested in positively influencing people to make better decisions at work. A health-related example of a 'nudge' would be the arrangement of healthy snacks at eye level in the workplace canteen. Other less healthy options, such as calorific chocolate bars, are given less visual prominence to ensure they aren't chosen as much.

In terms of safety, nudges can theoretically be used to prompt safe practice and actions, make the safest choice the default choice and increase awareness of our surroundings and hazards. Critics argue, however, that such attempts to subliminally influence behaviour tend to have a 'Big Brother' element to them. The effect of such nudges may also be short-lived, especially if they do not form part of a strategy designed to penetrate consciousness at a deeper level. Being outcome focused, the nudge approach pays less attention to the process of generating risk awareness. If unsafe behaviour is to be permanently changed for the better, it would make more sense to start with risk awareness before seeking out the mandate for behavioural change. Workers quickly see through management initiatives designed to manipulate their behaviour.

In order for any health or safety initiative to succeed, an appropriate level of buy-in is usually necessary. The nudge approach appears to want to influence by stealth and through the 'back door'. For this reason, it is unlikely to succeed in gaining much traction in complex environments, since to change people's behaviour we must be able to influence their minds. The nudge approach has relatively little to say here. It is perhaps best to think of it as a potentially useful but limited tool to be used judiciously in certain situations. Ultimately, however, it doesn't constitute a comprehensive enough framework for action within the health and safety domain.

Human Factors

The second half of the 20th century witnessed the growth of cognitive psychology, social psychology and engineering psychology. Mental and social phenomena entered the fray with these emerging disciplines. The old view that human beings were the cause of organisational trouble when something went wrong began to soften in some quarters. In its place, a more enlightened view began to gain credibility in stressing that human beings were just as likely to be on the receiving end of trouble. Researchers such as Fitts and Jones (1947), who influenced views on pilot error at the time, sowed the seeds of this. They said:

> It should be possible to eliminate a large proportion of so-called 'pilot error' incidents by designing equipment in accordance with human requirements.[16]

WWII pilots had been mixing up throttle, mixture and propeller controls. By opening up a window on their first-hand experiences, Fitts and Jones were able to determine that this was often because the locations of levers kept changing across different cockpits. This created a kind of mental interference, which, once identified, could be controlled with smarter, more person-centred design. This thinking has been capitalised on ever since, highlighting that systems should be designed to be as error resistant and error tolerant as possible. Human requirements should be placed at the heart of a system at the design stage, rather than as an afterthought.

The 1970s saw an increased focus on organisational factors going beyond the immediate working environments of frontline staff. Bureaucracy and management became the logical places to look for the original sources of industrial failures. The 'blunt end' rather than the 'sharp end' became the hunting ground, and managers were now in the cross hairs. The interest lay in upstream processes, such as management information gathering, decision-making and communications. The idea was that normal, bureaucratic processes could start brewing potential failures, long before anything became visible at the sharp end. Typified by Barry A. Turner (1978) in his book *Man-Made Disasters*, it set the stage for perhaps the most famous human factors model ever devised, which was named after a dairy product no less!

The Swiss Cheese Model

Do not be fooled by the name. Back in the 1990s, James Reason's Swiss Cheese Model quickly gained international credibility as a model of accident causation.[17] In Reason's model, depicted in Figure 2.1, an accident is seen as the result of many prior failures in organisational layers, or slices of cheese, which exist upstream of the sharp end where people work. For example, these layers consist of things like design, procedures, supervision and unsafe acts. To prevent accidents, we need to 'fix' the holes in the layers, which will serve as barriers against failures.

The Swiss Cheese Model might be visually pleasing, but its deterministic cause and effect thinking relies on the benefit of hindsight. This sits uneasily with the modern reality of workplace complexity. Its strength is the ability to communicate organisational failures effectively, but the model's enduring appeal disguises the fact that it offers little practical support once such failures are found. Just how do we plug these ill-defined holes? The tendency to highlight management failures may also eclipse important issues such as the impact of emotion on individual performance.

The next approach addresses some of the model's deficiencies by focusing more on positive events.

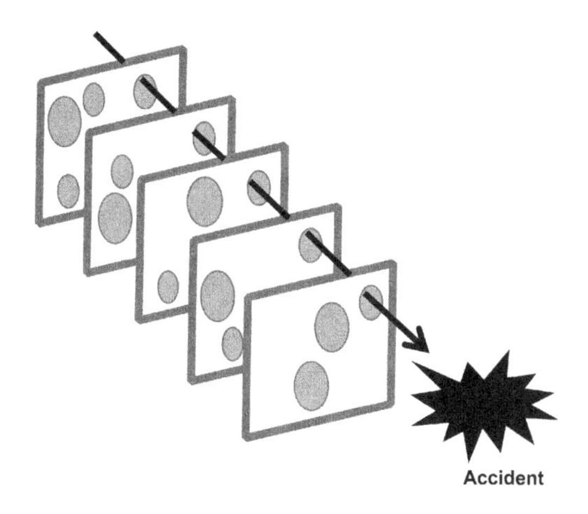

FIGURE 2.1 Reason's Swiss Cheese Model.

SAFETY-I AND SAFETY-II

Erik Hollnagel, a human factors expert, challenges the status quo regarding 21st-century views on safety management. His approach makes use of the concepts of Safety-I and Safety-II, which essentially represent different mindsets or philosophies.[18] In Safety-I, safety management is reactive and the goal is to achieve the lowest possible number of events going wrong. Most health and safety professionals would agree that safety should be about creating environments where there is an absence of harm or danger. A constant effort is needed to keep incidents to a minimum. Though this may be true, overemphasising learning from negative events can distort our perception of the reality of complex environments.

Hollnagel's central argument is a subtle, alluring one. In short, there is an abundance of information on what happens when things go wrong or fail, but this will only take us so far. Failures are only a tiny proportion of events overall, perhaps accounting for 1 in 10,000. In his thinking, we need to shift up a gear to achieve the mindset represented by a far more proactive Safety-II. Stuck in a reactive pattern of thinking, organisations typically focus on avoiding something going wrong, rather than ensuring that things go right.

In Safety-II, a special interest is taken in redressing the imbalance. Things usually do go right, and from these positive events, there is a huge, untapped reservoir of organisational learning. By understanding how things go right, we can provide positive examples of good safety practice for others to follow. We can also better explain how things occasionally go wrong. On the journey to Safety-II, Hollnagel suggests that there is a significant change in attitude to human error. In fact, the whole notion of human error might represent flawed thinking. The problem is that this concept often neglects the full situational context. Not only that, it is also historically overloaded with the baggage of finger pointing and blame. Frontline staff often find themselves in the firing line and the target of disciplinary proceedings.

The reality is far more complex, with management and frontline staff sharing responsibility when errors are made and things go wrong. Focusing less on human error, Hollnagel prefers to talk about 'performance variability'. Inevitably, there are differences between two individuals performing the same task. Also, an individual will perform differently depending on a range of factors, such as their age, experience, current workload and environment. This performance variability can be monitored and managed, as well as being a rich source of information.

Hollnagel's thinking has contributed much to the field of safety management. It provides a useful framework for analysing both positive and negative events, dispelling the unhelpful myths behind the conventional notion of human error. However, we should recognise that the analysis is almost exclusively confined to the level of safety management. To reflect on recent trends and developments, it is necessary to embrace the subject of health more whole-heartedly. This must also become part of effective safety management, since an absence of good health can cause accidents, as illustrated by the case of Germanwings 9525, which is discussed in Chapter 7. Neither does there appear to be much in the way of practical solutions to help achieve a Safety-II mindset.

This is where we must part company with Hollnagel's approach and embrace more holistic thinking. In the preceding sections, we have sufficiently set the scene by providing a review of the main approaches, but we must now turn our attention to the uniqueness of the M4 one.

TABLE OF BASIC DIFFERENCES

The M4 approach offers some considerable advantages over other approaches, as outlined in Table 2.1.

TABLE 2.1
The Basic Differences between Health and Safety Approaches

	Trains Focus & Concentration	Enhances Self-Awareness	Covers Mental & Physical Sides	Learning from Positives	Levels of Analysis[a]
M4	✓	✓	✓	✓	1, 2, 3, 4
Behavioural safety				✓	1, 3
Nudge					1, 3
Swiss Cheese Model					1, 3
Safety-I and -II				✓	1, 3

[a] Levels of analysis: 1 = Individual; 2 = Relational; 3 = Organisational; 4 = Societal.

SUMMARY OF M4 ADVANTAGES

Starting from Scratch

Fresh thinking is definitely called for here, and sometimes it is better to start from scratch. None of the approaches pay more than lip service to health, and neither do they cover the full range of levels: individual, relational, organisational and societal. The relational level very often appears to be forgotten completely.

Behavioural safety has its roots in the distant past, typically showing little tolerance for human beings with their own psychologies. Even its modern 'nudge' derivative suggests people can be controlled, or at least subjected to benign manipulation. The assumption that people are to be steered rather than mentally stimulated is flawed. Despite this recent resurgence, the general approach seems to have had its innings and is bereft of new ideas.

The human factors approach is seemingly far braver in its intention to explain people's behaviour, as well as the events they find themselves caught up in. However, though the pendulum may have swung from an emphasis on human error (individual level) to management failures (organisational level), the theorising in both cases leaves one a little cold.

A truly balanced approach would take full account of the overall context by evenly scrutinising events at all levels. Working selectively at the individual and organisational levels leaves too many gaps. In fairness, Hollnagel's recent work does usefully add the 'learning from positive events' dimension to the human factors approach, but this is still very much based around organisational theorising.

Making Sense of People's Experiences

In contrast, the M4 approach starts by valuing people as a source of experience, untapped learning and potential. They can provide health and safety insights and creativity. At all times, we therefore need to make a concerted effort to make sense of their experience. Rather than pigeonholing people's behaviours into safe or unsafe acts, we must embrace the full range of their on-the-job performance. We must accept performance variability, and learn from it, as it is a fact of life. In this approach, operational experience takes centre stage, and we want to remain as curious about it as is humanly possible. This also directs us to the key role played by risk awareness in safeguarding health whilst ensuring high levels of operational performance.

It is heartening to see writers such as Rosa Carrillo highlighting the importance of relationships in health and safety thinking. This is why relational mindfulness is as fundamental to the M4 approach as any of the other levels. It may have been largely forgotten, but its reinstatement here underlines the fact that healthy relationships create the right conditions for optimal safety performance. Open dialogue between frontline staff and their managers encourages revitalised conversations about health and safety.

Resisting the Status Quo

At the organisational level, we must resist some of the bureaucratic machinery of health and safety, with its focus on counting rule violations, deviations and incidents. When a self-serving bureaucracy prospers, paperwork can generate yet more paperwork, potentially at the expense of risk awareness and mitigation. People soon end up feeling disempowered, with thinking and initiative stifled. Collective inertia will impede any real progress in such circumstances. Organisational success in the field of health, wellbeing and safety can't be achieved by obsessively exploring failure and counting negative events.

Hollnagel has offered a partial solution to this intellectual cul-de-sac. Mindful organisations will be able to evaluate where they systematically focus their attention, structuring and articulating new priorities that promote learning from positive events. Nevertheless, cultural change may be difficult to effect across industries where there are liability concerns, and further regulation looms. Health, safety and wellbeing may be held hostage, especially where rules are put in place to deflect liability rather than prevent incidents.

Practical Applications

One of the biggest criticisms of the approaches surveyed is that they tend to offer little in the way of practical solutions. They might provide useful conceptual frameworks, but these are at best suggestive when it comes to tackling the underlying issues. The human factors approach has perhaps been the exception in providing a range of techniques for ergonomically improving physical environments. This is taken a step further with the M4 approach. For example, the loss of situational awareness described in many accident investigations can be greatly reduced with mindfulness training.

When, through regular practice, we learn to notice when our mind is wandering, concentration levels increase. Whether we are talking about flying, driving or operating machinery, the way we direct our attention is critical for successful, incident-free performance. Two such applications, where mindfulness was taught to London bus drivers and nuclear power stations staff, are discussed in Chapter 12, but there are many other applications. Mindfulness is also a key component of interpersonal communication, as when we openly attend to tone, words and non-verbal behaviour, we are far more likely to listen authentically, both to ourselves and to others. These skills are often neglected in traditional schooling, but they can be taught in the workplace. Personal care and looking after ourselves with compassion are central tenets of mindfulness practice (see Chapter 4). Before safety, there is health, and the mental and physical health of frontline staff is hugely important for those who must perform safety critical tasks. It is equally important for managers, who are expected to make good decisions that impact on their staff. A stressed-out workforce, driven hard to meet production targets, is unlikely to provide a sound basis for a positive health and safety culture. Mindfulness provides the mental 'toolkit' to respond effectively to low mood, with proven success in preventing a spiral into depression. Mindful organisations will want to cultivate mindful individuals in order to increase happiness and productivity levels. This doesn't mean there should be a carte blanche for the creation of more compliant, socially engineered workers. Organisations adopting this way of thinking will likely edge instead towards challenging the existing order, with implications for society too.

Once we put people's experiences at the heart of an approach, a revolutionary new way of looking at things is opened up. We are less interested in hindsight and attempting to fix things once they have gone wrong and more interested in helping them go right in the first place. We need to be less judgemental of ourselves and others. When errors are made, we need to be more forgiving, and refrain from taking an unfavourable view without a full appreciation of the circumstances. This works just as well at an organisational level too. This is why listening to frontline staff and their concerns, well before there is an incident of some kind, is so important here.

THE M4 CALL TO ACTION

The M4 approach is a call to action for increasing our own mindfulness in the first instance, listening more to others and systematically paying attention to the things that matter. This increases self-awareness and risk awareness, promotes wellbeing, leads to more fulfilling relationships and improves safety performance. It also has societal benefits and marks a clear break with past approaches, which tend to be short on practical applications and mired in static, purely analytical thinking.

In contrast to most approaches, learning is applied across four levels instead of just one or two. It requires us to attend to ourselves, our colleagues and our surroundings in a very practical way. If we want to make real gains in health, wellbeing and safety, and prevent incidents with a societal cost, we will need to think differently.

By demonstrating mindful leadership, you could be part of the change in your own organisation. Contact the M4 Initiative for details: www.m4initiative.com

NOTES

1. Kabat-Zinn, J. (2004). *Wherever You Go, There You Are. Mindfulness Meditations for Everyday Life*, p. 17. London: Piatkus Books.
2. LeBeau, P. (2018). Traffic deaths edge lower, but 2017 stats paint worrisome picture. *CNBC*. 15 February 2018. Available from https://www.cnbc.com/2018/02/14/traffic-deaths-edge-lower-but-2017-stats-paint-worrisome-picture.html (Accessed 01.08.2018).
3. Distracted Driving (2015). Traffic safety facts. Research note. National Highway Safety Traffic Administration. Available from https://crashstats.nhtsa.dot.gov/Api/Public/ViewPublication/812381 (Accessed 01.08.2018).
4. Carrillo, R.A. (2012). Relationship-based safety. Moving beyond culture and behaviour. *Safety Management*, Volume 57, pp. 35–45.
5. Gittell, J.H. (2009). *High Performance Healthcare: Using the Power of Relationships to Achieve Quality, Efficiency and Resilience*. New York: McGraw-Hill.
6. Surgeons' 'toxic' rows added to mortality rate, says report. *BBC News*. 4 August 2018. Available from https://www.bbc.co.uk/news/uk-45067747 (Accessed 04.08.2018).
7. Gittell, J.H. (2003). *The Southwest Airlines Way: Using the Power of Relationships to Achieve Quality, Efficiency and Resilience*. New York: McGraw-Hill.
8. Simard, M. & Manchard, A. (1995). A multilevel analysis of organizational factors related to the taking of safety initiatives by work groups. *Safety Science*, Volume 21, pp. 113–129.
9. National Commission on the BP Deepwater Horizon Oil Spill and Offshore Drilling (2011). *Deep Water: The Gulf Oil Disaster and the Future of Offshore Drilling*, pp. 2–6. Available from https://www.gpo.gov/fdsys/pkg/GPO-OILCOMMISSION/pdf/GPO-OILCOMMISSION.pdf (Accessed 01.08.2018).
10. Ibid., p. 90.

11. Chernoff, A. (2010). Study: Deepwater Horizon workers were afraid to report safety issues. *CNN news.* Available from http://edition.cnn.com/2010/US/07/22/gulf.oil.rig.safety/index.html (Accessed 07.11.2020).
12. Weick, K. & Sutcliffe, K. (2001). *Managing the Unexpected. Assuring High Performance in an Age of Complexity*, p. 3. San Francisco, CA: Jossey-Bass.
13. McInerney, P. (2001). *Final Report of the Special Commission of Inquiry into the Glenbrook Rail Accident.* Sydney: NSW Government.
14. Lingard, H. & Rowlinson, S. (1994). Construction site safety in Hong Kong. *Construction Management and Economics*, Volume 12(6), pp. 501–510.
15. Thaler, R.H. & Sunstein, C.R. (2008). *Nudge. Improving Decisions about Health, Wealth and Happiness.* New Haven, CT and London: Yale University Press.
16. Fitts, P.M. & Jones, R.E. (1947). *Analysis of Factors Contributing to 460 "Pilot Error" Experiences in Operating Aircraft Controls.* Dayton, OH: Aero Medical Laboratory, Air Material Command, Wright-Patterson Air Force Base.
17. Reason, J. (1990). *Human Error.* Cambridge, UK: Cambridge University Press.
18. Hollnagel, E. (2014). *Safety and Safety-II. The Past and Future of Safety Management.* Farnham: Ashgate.

3 New Tools for Incident Investigation

> In the field of human relations nothing is so important as safety, for safety applies with equal force to the individual, to the family, to the employer, to the country. Safety in its widest sense, concerns the happiness, contentment and freedom of everyone.
>
> – **Bill Jeffers**

The M4 approach can be applied very effectively to accidents, in order to explain rather than judge how events unfolded and led to catastrophic consequences. In contrast to the approach taken in most accident investigations, the emphasis is on empathically reconstructing events for all those who played a role at the time. In reality, the reconstruction of complex environments after the event is fraught with difficulty, but this doesn't prevent us from learning lessons.

This chapter presents a working case study of Air Florida Flight 90, where you will see the analytical side of the M4 approach in action. Despite the obvious tendency to blame human error for the accident, many other factors were responsible. By analysing things from each level in turn, we can obtain a more balanced perspective and address the systemic failings too. This is an approach we will be taking throughout the book.

Case Study: Air Florida Flight 90

This case study of Air Florida Flight 90 applies M4 thinking to a major disaster. There is evidence of mindlessness occurring at all levels here. In an effort to identify issues and circumstances that we may all face in the workplace, we are interested in understanding what happened at a deeper, experiential level. Blame must always be jettisoned if we are to learn from the 'errors' of the past. With the benefit of hindsight, it is far too easy to label human actions as errors, and far too easy to ignore the full situational context. If the highly trained professionals often held responsible for so many accidents could turn back time and make a different set of decisions, they surely would. Taking a more holistic approach, we must also free our thinking from the analytic reduction of Newtonian cause and effect.

Flying Mindlessly

When things go wrong for pilots, the consequences can cause loss of life on a horrific scale. The importance of being mindful at all levels is illustrated well by the example of Air Florida Flight 90. Picture a very wintry day at Washington National Airport.

It is 13 January 1982, and the temperature is 4°C. When Air Florida Flight 90 finally accelerates down the runway for takeoff, just before 4 pm, heavy snow is falling. There has been a 49-minute wait in a taxi line with many other aircraft. In the cockpit are the 34-year-old captain, Larry Wheaton, and the 31-year-old first officer, Roger Petit.

Just 90 seconds later, the Boeing 737, in an extreme nose up pitch attitude, falls out of the sky and crashes into the 14th Street Bridge over the Potomac River. Before plunging through the ice into the water, the aircraft strikes seven vehicles on the bridge, killing four motorists. In total, the aircraft was carrying 74 passengers and five crew members. Tragically, just four passengers and one flight attendant survived after being rescued from the freezing river. What went wrong?

The subsequent investigation by the National Safety Transportation Board (NSTB) pointed, amongst other things, to the flight crew's pre-takeoff control checks in the cockpit.[1] The first officer dutifully called out each control on the list. The captain and first officer were meant to ensure that the switches were in the correct positions. This is the conversation as it happened:

First officer: Pilot heat?
Captain: On.
First officer: Engine anti-ice?
Captain: Off

Remember that it was freezing outside. Though they ran through the checklist together, the engine anti-ice switch was never moved into the 'on' position. The flight crew were acting mindlessly. The captain and first officer were just going through their routine checklist, just as they had always done. This time, however, they were not flying in the weather they were accustomed to. It was icy and the conditions were treacherous. It is not difficult to empathise with the flight crew under these circumstances. These pre-takeoff control checks are similar to the safety demonstrations given by flight attendants. As passengers, we may often find these tiresome. The consequence of this is that we may 'tune out' and ignore important safety information. Blindly following routines can make us behave like automatons. Our eyes glaze over and our attention falls away.

Assumptions Fill the Void

The flight crew were aware that some snow or ice had accumulated on the wings, and in the cockpit, they discussed deicing at length. They even made an attempt to clear the wings themselves, by positioning themselves closely behind another aircraft for additional heat, assuming that this might help. Assuming, rather than thinking, is one of the obstacles to attaining mindfulness. Instead, the exhaust gases from the other aircraft probably turned the snow into a slushy mixture, liable to freeze on the wing leading edges. If it hadn't been exposed to those exhaust gases, the snow would likely have been blown off during takeoff. The flight crew's decision to break with the standard procedure of maintaining distance between aircraft must, however, be understood in the wider context.

It was known that the B-737 they were flying could 'pitch up' when the wing leading edges were contaminated with even small amounts of snow or ice. This could seriously affect the aircraft's aerodynamic drag. In simple terms, changes in the contour shape of the wing, caused by snow or ice, could significantly affect the airflow, leading to a reduction in lift. The general effectiveness of the deicing procedures used to protect aircraft in such conditions had not been ascertained for prolonged waiting periods. This left the flight crew guessing, and they filled the information void with their own assumptions, not really knowing how effective the deicing procedure had been back at the gate. In the accident report, the NSTB expressed their concern that:

> ...pilots may erroneously believe that there is a positive protection provided for a period following the application of deicing/anti-icing solution, which eliminates the need to closely monitor the aircraft for contaminants during ground and takeoff operations.[2]

Whatever they really thought about the degree of protection afforded to them, the flight crew found themselves in a situation not entirely of their own making. They were positioned downstream of various mistakes, information voids and communication lapses made by others responsible for the deicing procedure back at the gate.

Falling between Stools

At the gate, American Airlines' maintenance personnel had deiced Flight 90 under an existing service agreement. American Airlines did not actually operate any B-737 aircraft themselves. Surprisingly, there had been little communication between the two companies about the procedures that should be used to deice this specific aircraft.

The Air Florida Maintenance Manual included little information in relation to deicing, but it did stipulate that covers for static ports (used for air instrumentation data), and plugs for engine inlets, should be in place when deicing fluid was applied. American Airlines did not comply with this requirement. In the absence of proper discussions between the two companies, the maintenance responsibilities were not defined or fully understood. This set the stage for further mishaps.

In the midst of this organisational confusion, the maintenance operators responsible for deicing Flight 90 were left in ambiguous territory. The left side of the aircraft was deiced first. The operator selected a mixture of approximately 35 per cent deicer and 65 per cent water, applying a final overspray at the same time. This operator was then relieved by another who tackled the right side of the aircraft using 100 per cent water followed by a final overspray, applied with approximately 25 per cent deicer and 75 per cent water.

Tests later showed that the mixture dispensed was far more diluted than the selected one. Only 18 per cent of deicing fluid was in the mixture, rather than the 30 or so per cent intended. The inaccurate mixture was attributable to a replacement nozzle on the delivery hose, which was not calibrated in the same way as the original one. A mix monitor was not available on the Trump deicing vehicle used, though these were installed on the most up-to-date vehicles. The mix monitor would have ensured greater accuracy by providing a visual reading.

Before the aircraft left the gate, no witnesses could recall seeing either the captain or the first officer leave the cockpit to check for any remaining snow or ice. A visual inspection by the flight crew may have increased their awareness of the conditions and the risks they faced. It is difficult to say if this would ultimately have made any substantive difference to the way events unfolded.

The inadequate state of communications between Air Florida and American Airlines certainly did not help. M4 thinking extends the use of mindfulness to cover the influence of communications and relationships, or 'relational mindfulness'. Many individuals and teams are responsible for an aircraft's safe takeoff, and the way they communicate with each other is critical. The quality of the communication is always important, as is knowing when to speak up to prevent disaster, especially in the cockpit. We will turn now to the critical conversation between the captain and the first officer just before takeoff.

Challenging Authority

The engines spooled up. Soon after the takeoff roll began, the first officer interjected several times and warned that the Engine Pressure Ratio (EPR) thrust gauges appeared to be grossly incorrect. His words were, "God, look at that thing. That doesn't seem right, does it?" The gauges were in fact providing false readings because of the ice-blocked probes. Remember that there had been a failure to activate the engine's anti-ice systems. Lower-than-normal engine thrust settings were hampering acceleration down the runway, but the captain dismissed these concerns. The takeoff roll went ahead.

Challenging the captain's authority during the takeoff roll required more assertiveness. The first officer did in fact express concern that something was 'not right' four times, but the captain took no action to reject the takeoff, knowing that he alone made the decision to reject as stipulated by procedure. The investigators said:

> With regard to the first officer, while he clearly expressed his view that something was not right during the takeoff roll, his comments were not assertive. Had he been more assertive in stating his opinion that the takeoff should be rejected, the captain might have been prompted to take positive action.[3]

Note that it was not a lack of communication that was said to contribute to the unfolding disaster, but the first officer's unassertive tone. Communication is more than just the exchange of information between two individuals. The first officer opened his mouth and words came out, and the captain was the target of those words. The captain signalled that he had heard them, yet he ignored the gravity of the situation and failed to appreciate the risk.

It is entirely possible that the captain was not receptive to his first officer's interjections because he felt his professional or social status was being undermined. He might well have been acting to protect his position and authority. By doing this, his range of options to avert disaster became narrowed. In everyday situations, the consequences aren't usually catastrophic if people protect their social statuses in their communications with each other. It is very different in the cockpit of a 46-tonne aircraft. Learning to respond rather than react to what another person says is a key part of relational mindfulness. If we are carried away by our reactions and the psychological defences we employ unthinkingly, our thinking becomes blinkered and our responses limited.

Beyond Hearing

When someone speaks, relational mindfulness requires that the other person should listen and not just hear. To be attuned to the importance of what the first officer was conveying in his communication, the captain had to be in a receptive frame of mind. A fundamental prerequisite for authentic listening had not been met, possibly because the captain was protecting his social status or was distracted. They had been told by the local air traffic controller not to delay, as there was already another aircraft on the final approach to the same runway. Calculations by investigators later clarified that, in theory at least, there had been enough time to successfully reject the takeoff, but the captain probably felt that there was no more time to lose.

With the benefit of hindsight, not rejecting the takeoff was a failure to listen to the first officer, or just poor judgement. That view, in failing to acknowledge the inner pressure the captain was under, might lack compassion. In his mind, it was the right decision at the time. He hadn't heard a strong enough challenge to stop the aircraft accelerating down the runway, and the words of the local air traffic controller might have been echoing in his mind. Whatever the case, his mind was closed to the option of rejecting the takeoff. Anxious to reach their destinations, the flight crew and 74 passengers had waited long enough.

It was a rough ride down the runway. Once airborne, the stickshaker, the instrument that warns of an imminent stall, began to vibrate. It would continue to vibrate right through to impact. The first officer's last words were, "Larry, we're going down ..." to which the captain replied, "I know it." The aircraft was airborne for just 30 seconds, reaching a maximum altitude of 352 ft.

Being Last Is Painful

After leaving the gate, the aircraft had to wait 49 minutes in a queue of many other aircraft before reaching the runway for takeoff. In this time, there was a slow accumulation of more snow and ice on the wings. In hindsight, the pilot's decision not to take the aircraft back to the gate for a second deicing application was catastrophic. It looks like an incredibly poor decision, at least at face value.

Yet there is more than meets the eye going on here. Our minds are strongly affected by the queuing process. After the wait, think of the overwhelming pressure to get the aircraft off the ground. How many pilots would have given up a hard-earned place and decided to go to the back of the queue? No one likes to find themselves stuck at the end of a line. The NTSB accident report recognised all the time that would have been lost in dropping out of the queue:

> There is not sufficient room under most circumstances to get out of line and taxi to a designated area for deicing and then fall back in line for takeoff. The flight crew's options are limited to continued waiting until they are able to takeoff, or returning to their deicing areas, where they will probably be exposed to more waiting for space at the ramp.[4]

It has been argued by economists at Harvard Business School that 'last place aversion' might be an innate human trait, or a conditioned response. Whatever the truth of the matter, the phenomena's effects can be observed in society at large: in traffic, grocery store and call centre queues everywhere. And, of course, in aircraft queues for takeoff. Shame or embarrassment may keep individuals from doing anything that puts them in last place.[5] To avoid being last, they will often gamble and make riskier choices, and in the case of Air Florida Flight 90, this is a very plausible factor. Whether or not the pilot and co-pilot were conscious of their last place aversion, it was far riskier to takeoff under those conditions than to turn back for the gate.

Mentally Pulling Out of Queues

Think about a scenario where you are turning out of a side road onto a busy main road in your car. You have a passenger with you and you are both heading to work. You are trying to make the turn across one lane of traffic, but there is no break in the constant stream of cars. No one is kind enough to stop and permit you to turn. Directly behind you, a queue of cars has built up, and both you and your passenger are anxious to get to the office on time. The motorists queuing behind are becoming increasingly frustrated. One has already tooted their horn. Two minutes later, you are tempted to take a risk and make the turn in front of an oncoming car that is travelling a little too fast for the conditions.

I'm not suggesting that a viable option would be to pull out of the traffic queue altogether. In the case of Air Florida Flight 90, leaving the queue to return to the gate was absolutely necessary to prevent disaster. To avoid increasing the risk of a road traffic accident, however, it is necessary to 'mentally pull out of the queue'. Mindful of the pressure, the best course of action is to calm your nerves in order to respond rather than react to the situation. This is precisely where mindfulness training is indispensable. Being aware of our reactions to stressful situations, whether they manifest themselves in bodily sensations, or the kinds of thoughts we have, is the key to choosing the safest course of action.

A Tolerance of Danger

The flight crew were aware before takeoff that the tops of the wings were covered in snow or slush. They apparently did not believe this would significantly affect the takeoff, or the ability of the aircraft to climb. Neither did they want to forego the takeoff opportunity for another round of deicing, and further delay the flight unnecessarily. They could see that other aircraft were taking off successfully without incident. All the flight training material on winter operations stressed the importance of 'clean' wings for takeoff. At that time, however, a form of collective mindlessness existed amongst pilots at a professional level. They tended to believe that snow or ice on the wings did not pose a significant risk. The flight crew were certainly not alone in their beliefs, as highlighted by the NTSB:

> ...the Safety Board believes that this crew's decision to take off with snow adhering to the aircraft is not an isolated incident, but is a too frequent occurrence.[6]

This under-appreciation of the risks involved was probably reinforced by pilots' common experience of icing during cruise flight, which was normally encountered without difficulty. Flight manual statements also suggested that aircraft could cope perfectly with icing, once cruise altitude had been reached. This is a good example of how ill-informed assumptions can create the conditions for a kind of mindlessness prevailing at all levels, from the individual level through to the organisational and beyond. A tolerance of danger thrives where the safety risks are misunderstood or under-appreciated, and in the absence of scientific research.

Winter Operations

The NTSB concluded that the flight crew had shown insufficient concern for the winter hazards they faced.[7] An important question to ask is why? After upgrading to his role flying B-737 aircraft, the captain had only flown eight takeoffs or landings in similar freezing or near-freezing conditions. The first officer had only flown two takeoffs or landings in such conditions after joining Air Florida. Both young and in their 30s, the risk of their combined inexperience in such conditions could have been mitigated with more robust training. The Air Florida training regimen covered cold weather operating procedures in the classroom, but there was a notable omission. It did not include detailed discussions of the possible effects on instrument readings if the engine anti-ice system was off. Knowledge gained in this area may have made the captain more receptive to the possibility of faulty thrust readings. He would then have possessed the mental tools to make a more effective diagnosis of the situation he faced on the runway.

Neither did the formal training provide sufficient opportunities to demonstrate cold weather operations under the guidance of an instructor. It is one thing to be shown presentations, films and slides, and be given lectures in the classroom, but quite another to practise with the benefit of expert feedback from an instructor. Knowledge of cold weather operations was far more difficult to operationalise, retain and commit to memory under these conditions. In the case of Air Florida, important operational knowledge for dealing with winter hazards never made it out of the classroom.

The Training Dilemma

This brings to the fore an ever-present training dilemma: how do we effectively train operational staff to respond appropriately to a scenario they will encounter infrequently? When the scenario finally arises in real life, the knowledge of how best to tackle it may have long faded from memory. And that is if the knowledge required to respond has been committed to memory in the first place. It is a bit like driving a car on an icy road and finding yourself in a skid. You may not encounter these road conditions often, yet you need to be prepared for them when they happen. Your immediate response upon detecting a skid will determine whether the car stays on the road or ends up communing with the roadside shrubbery. Skid control is not something routinely taught by most driving instructors either. Ideally, you would want to give drivers live instruction in this under safe conditions, away from real traffic. This form of training then needs to be refreshed at regular intervals.

In short, it is important not to analyse human error in an isolated context. The errors made by Flight 90's flight crew took place downstream of a limited organisational training regimen, which failed to instil some of the basics for winter operations. It can be deduced that organisational mindfulness plays an equal role to individual mindfulness in the prevention of accidents.

M4: APPLYING MINDFUL SAFETY TO AIR FLORIDA FLIGHT 90

The case study of Flight 90 shows how the analysis of complex accidents must be approached from a multi-level perspective. We can achieve a much greater understanding of accident causation by looking at the contribution each level makes to the big picture, empathically recreating the events to gain more insight. There are many lessons that can be learned from Flight 90, and these can usefully be applied in other domains where safety is of paramount importance.

Individual

- We must always be alert to the dangers of mindlessness, as it can cause huge problems when it is allowed to develop in safety critical situations. A clear example is provided by the case of Flight 90 and the tragic loss of life.
- Pilot training in mindfulness can prevent the loss of awareness being played out in tragic circumstances.
- Checklists provide an opportunity for mindlessness, if they are carried out routinely through force of habit.
- In the absence of evidence and research, we can be prone to fill the void with assumptions, which may appear logical. Though they may be extrapolated from other more familiar situations, they can be deadly if they are applied inappropriately in safety critical situations.

Relational

- We may think we are communicating effectively, but communication is more than just an exchange of words or information.
- The captain of Flight 90 heard the words spoken by his first officer, but he did not respond appropriately.
- To avoid narrowing our range of options, we need to cultivate our ability to respond rather than react to what another person says.

- Our professional or social status can be a hindrance in communications. If we act to defend it, we may find ourselves unable to listen. In safety critical situations, this may threaten life.

Organisational

- If two organisations fail to communicate effectively over contractual arrangements and their practicalities, it can cause confusion and create significant safety risks.
- Clear instruction manuals to help carry out maintenance activities are essential, especially where aircraft with unique characteristics are concerned.
- Pilot training for winter operations could have been improved at Air Florida. Insufficient attention was given to operationalising knowledge gained in the classroom.
- A training dilemma is posed wherever operational staff have to meet the challenge of a situation that arises infrequently. Their skills need to be maintained for such situations.
- A professional tolerance of danger can thrive where safety risks are underappreciated or misunderstood. Awareness raising activities can help counter the attitudes that drive unsafe behaviours.

Societal

- Being aware of societal pressures is also a key component of mindfulness. Societal values, attitudes and patterns of behaviour can be hugely influential in our thinking.
- 'Last place aversion' is a good example of how our behaviour can be determined by collectively held views in wider society. We can take riskier decisions to avoid being last, as borne out by the research.
- Dropping out of the queue for takeoff was improbable for Flight 90, but staying increased the risk of an accident massively.
- We can experience the same kind of pressure in road traffic queues.
- Societal pressures need to be resisted to reduce safety risks in certain situations. Being able to 'mentally pull out' of a queue and resist these pressures can save lives.

NOTES

1. National Safety Transportation Board (1982). *Aircraft Accident Report. Air Florida, Inc. Boeing 737-222, N62AF.* Collision with 14th Street Bridge, Near Washington National Airport, Washington, D.C. January 13, 1982, p. 7 of transcript. Available from http://libraryonline.erau.edu/online-full-text/ntsb/aircraft-accident-reports/AAR82–08.pdf (Accessed02.08.2018).
2. Ibid., p. 56.
3. Ibid., p. 68.

4. Ibid., p. 69.
5. Kuziemko, I., Buell, R.W., Reich, T. & Norton, M.I. (2011). "Last Place Aversion": evidence and redistributive implications. *NBER Working Paper* No. 17234, July 2011.
6. National Safety Transportation Board (1982). *Aircraft Accident Report. Air Florida, Inc. Boeing 737-222, N62AF.* Collision with 14th Street Bridge, Near Washington National Airport, Washington, D.C. January 13, 1982, p. 62. Available from http://libraryonline. erau.edu/online-full-text/ntsb/aircraft-accident-reports/AAR82–08.pdf (Accessed 02.08.2018).
7. Ibid., p. 66.

4 Self-Care
The Cornerstone of Mindful Safety

If your compassion does not include yourself, it is incomplete

– **Buddha**[1]

Having described in detail the four different levels of mindfulness in Chapter 2, this chapter is mostly about how we can look after ourselves more compassionately, both at home and in the workplace. Though this is obviously not a self-help book, we are going to be taking more than a cursory glance at some of the changes we could be making. Lifestyle factors such as sleep, diet, exercise and being present are all areas that deserve our attention for potential improvement. Making these often entails the breaking of habits that are not conducive to our wellbeing or safety.

We'll begin by taking a look at some of the known health benefits of practising mindfulness, which I hope will provide some motivation for making those changes. The great thing about mindfulness is that it puts the same emphasis on the physical and mental side of things, treating both as part of a system.

It also provides the perfect toolkit for breaking unhelpful habits and learning more positive ones. And in the age of Covid-19, this is of critical importance if we are all to stay safe and well.

THE HEALTH BENEFITS OF MINDFULNESS

There are some well-researched health benefits to practising mindfulness, or mindfulness meditation as it is sometimes called.

Benefits of Regular Practice

- Studies show that meditation can bolster the immune system. This plays a role in helping to fight off colds, flu and viruses.[2]
- Even for more serious conditions, such as chronic pain,[3] cancer[4] and drug and alcohol dependence,[5] meditation has been found to work effectively.
- Anxiety, depression and irritability can all be decreased with regular meditation.[6]
- People who meditate are happier, on average, than those who do not.[7]
- Meditators tend to experience more fulfilling relationships, because communication is improved through greater empathy.[8]

One of the most incredible findings from the research is that the brain physically changes with mindfulness meditation practice. These changes can be seen in colourful brain scans after just eight weeks. We are clearly not just practising to create rainbow-like brain scans, but this offers hard proof that mindfulness works at the neurological level!

The brain regions responsible for learning, memory, emotional regulation and empathy grow thicker. In stark contrast, one critical area of the brain actually gets smaller. This is good news, because it is the amygdala, the almond-shaped 'fight-or-flight' centre responsible for anxiety, fear and stress.[9]

Although most of the health benefits are associated with regular practice, there is an easy way to get started with mindfulness meditation. Everyday activities can be used as a way of breaking out of autopilot and being present in the moment.

EVERYDAY MINDFULNESS AND 'BEING PRESENT'

When we are caught up in the midst of daily routine and work activities, we may not be fully awake to what is happening right now in front of our eyes. It can be fruitful to pick a few activities that appear to be part of the daily grind and turn them into 'mindfulness bells'. These bells can be used as reminders to stop and attend to the moment.

Here are some suggested activities:

Brushing your teeth. Are you on autopilot when you attend to your pearly whites? There are plenty of sensations to pay attention to: the flavour of the toothpaste, the feel of the brush on your teeth, the moisture in your mouth and any sensations in the wrist as you work your way around the mouth.

Taking a shower. Many people spend some of their time in the shower reflecting on things, or they plan what they need to do next. If you use your time this way, do so intentionally so that you are aware of where you are purposefully focusing your attention. Showering can, however, provide plenty of sensations without engaging in any additional mental activity. The feeling of water on your hair and body, the water's temperature, and the muscle movements required to apply shampoo or soap are all part of the experience.

Preparing food. While many people see food preparation as a chore, it is also a great opportunity to bring a quality of awareness into the frame. What does it feel like to handle food with your hands, or to chop vegetables with a knife? Try focusing on the process of cutting a carrot into slices of even thickness. Are you fully present for these activities?

Washing dishes. Seemingly, this is one of the most boring chores, and it can potentially be a big source of domestic arguments. However, it can be reframed as the perfect excuse for greater mindfulness. Explore the sensations evoked by running water, paying particular attention to its flow and temperature.

Driving. Driving provides a rich landscape of moving objects, sights and sounds. If our minds wander and too little attention is paid to the road, it can cause an accident. It is therefore important to train the mind to acknowledge any decisions that are made that involve shifting your primary focus away from driving. If, for example, you shift your attention to a forthcoming meeting you are attending, become conscious that this is a decision you are making. Then closely observe how it affects your driving performance. Can you shift your focus back again fast enough to attend to a driving situation that requires close monitoring, assessment or action?

If there is a passenger in the vehicle with you, be mindful of how any conversation impacts your ability to concentrate. This will likely vary from individual to individual. It may be possible to listen fully without the conversation interfering with your performance on the road. However, thinking about what to say next, or how to construct a response, may place an attentional load on you as a driver. It is for you to judge how this affects your decision-making ability, and ultimately your safety. Appropriate adjustments to your driving may need to be made. On occasion, this may mean politely disengaging from the conversation so you can concentrate more completely on the road.

Driving comprises many activities. For example, scanning the road ahead, looking in the mirrors and shifting your vision from close up to far away. These are all worthy of your attention. There are also many 'micro-activities' involved in driving, which can easily be forgotten about. These include touching the steering wheel (or the indicator stalk) and pushing on the pedals to accelerate or brake. Mindful driving can bring these micro-activities back into awareness when there is an opportune moment.

There are plenty of environmental cues that can prompt us to pause whilst driving and then focus on our breathing. Red lights, stop signs and traffic queues can all provide a little breathing space. Even a single, deep breath can give us some respite from driving.

Walking. We usually walk on autopilot. This is why this activity lends itself so well to mindfulness practice. Paying attention to the actual sensations of walking is the aim here. Notice when the mind begins to wander off and bring it back to 'just walking'.

TRADING SLEEP FOR LEISURE TIME

To be fully present in our everyday activities, we are going to have to get our lifestyle basics right. Are you reading this bleary-eyed? There's a good chance you are. The average Briton now sleeps for just six-and-a-half hours a night,[10] according to the aptly named Sleep Council, who definitely know a thing or two about nodding off.

Our lifestyles are contributing to a formidable 'sleep debt' across the nation, which has a direct impact on our ability to do our jobs safely and to the best of our ability.

If you are not getting the recommended seven to eight hours' sleep a night, you are probably not at your most alert at work. That will inevitably affect vigilance and the ability to make safe decisions, even for a routine activity such as crossing the road.

These days, it is a cultural habit to trade sleep for leisure time – everyone's doing it. We can squeeze additional leisure time out of our busy schedules in myriad ways:

slouched in front of the TV, with a pint in our hand at the pub, or crunching abdominals in the gym. But it is likely to be at the expense of time well spent in the land of nod. The CEO of Netflix has ominously remarked that the company's biggest competitor is not Amazon Video or YouTube, but sleep.[11] If only the day could be extended by a few hours to shoehorn a bit more leisure in.

An Hour's Extra Sleep Can Make All the Difference

Sleeping just an hour less than the recommended amount each night can actually impact our health at the genetic level. The University of Surrey carried out some research into this,[12] comparing a randomly allocated group of volunteers receiving six-and-a-half hours' sleep with another receiving seven-and-a-half hours' sleep over a period of a week. At the end of that week, the groups were asked to switch sleep patterns. Blood tests revealed that around 500 genes were switched on or off by changes in sleep patterns. Most worryingly, there were increases in the activity of genes associated with heart disease, diabetes and the risk of cancer.

More frequent Saturday and Sunday lie-ins might appear to be the solution, but burning the candle at both ends and then attempting to erase the sleep debt at the weekend is not the answer either. Important night-time brain processing, such as the consolidation of memories, needs to happen within 24 hours of the memories being formed, otherwise they may be lost. An hour of extra sleep makes all the difference and also improves cognitive functioning the next day. Try this during the week rather than attempting to compensate at the end of it – you'll feel better at work too.

The 28-Hour Day

Just imagine what you could do with an extra four hours at your disposal every day. A fascinating experiment carried out in 1938 by the so-called 'Father of Sleep Science', Dr Kleitman, who later went on to discover the rapid eye movement phase of sleep, aimed to test whether the human body could adapt to a 28-hour day.[13] Dr Kleitman and a fellow researcher essentially isolated themselves underground for a month in Mammoth Cave, in Kentucky. But adapting to the new routine in the absence of the usual environmental cues proved incredibly difficult for the cave dwellers, serving to remind us how tied to the 24-hour clock our bodies really are. We can't break free from thousands of years of evolution simply by staying up later, without it affecting our health.

But what do we do when it is 11 pm and we are hooked on a late-night movie? The typical response is to cut back on our sleep, opting for more leisure time. Perhaps because it is a personal choice for so many, and reinforced by popular culture, employers have been reluctant to tackle the lack of sleep amongst their workforce. The current approach to tackling the symptoms of fatigue generated by sleep debt is more about reacting once the horse has bolted than tackling the root cause. It doesn't help that the symptoms of fatigue are often as invisible as they are pervasive, and they can't be detected with a blood test or breathalyser. The best way forward would be to encourage everyone to get an extra hour of sleep every night, sacrificing some leisure time in the process. All the evidence suggests that it would significantly reduce human error and the number of accidents.

Sleep Your Way to Better Performance

The amount and quality of the sleep we get is vitally important. It underpins our mood and has a profound effect on our work performance. In a major study on elite performers in the fields of sport, arts and sciences, a key finding was that the best performers slept for eight hours and 36 minutes on average.[14] Compare that to the average six hours and 30 minutes that people usually get a night. In a sleep-deprived nation, our chances of Olympic-like performance in a variety of fields are being curtailed, often to the detriment of safety. Losing 90 minutes of sleep can reduce alertness by a third. Consider the potential consequences for doctors, pilots, drivers and construction workers, who all manage the safety and lives of others on a daily basis. Provided reasonable working hours and rest periods are adhered to, sleep is an area we can regain some control over. It is also a very measurable feature of our lives – sleep analytics are available through popular smartphone apps for just this purpose.

Bodies That Tell the Time

Our internal body clock, or circadian rhythm, tells us when to wake up and when to feel sleepy. We are beginning to understand the genetic mechanism by which this works in far more detail. In 2017, the Nobel Prize in Physiology or Medicine was won by three US scientists, who decoded how our bodies tell the time based on their work on fruit flies. The scientists showed how this works at the cellular level.[15] A metaphorical clock is ticking away in nearly every cell of the human body, influencing our mood, hormone levels, temperature and metabolism. These clocks are in evidence in animals, plants and fungi too. Disrupting our body clock, which effectively controls how our body matches night and day, can have a profound effect on our cognitive performance, leaving us feeling 'jet-lagged'.

Healing Faster

Our heart cells have circadian rhythms, just like other cells. There is even evidence that our chances of surviving heart surgery are significantly better in the afternoon.[16] You might be wondering if this is because heart surgeons, being human, are simply groggier in the mornings, but this effect was accounted for with the scientific controls in the research. Operations such as heart valve replacements require stopping the heart, placing it under considerable stress. In afternoon surgery, just nine per cent of patients suffered an adverse event, as opposed to a significantly higher 19 per cent for morning surgery.

Another truly incredible finding came from a study of 118 burns patients.[17] Burns sustained in the daytime took an average of 17 days to heal. Patients who were burned after dark took an additional 11 days to heal. Once again, we must go down to cellular level to explain this. In this case, skin cells called fibroblasts, the body's first responders, reacted far quicker during the day than at night.

The Exhaustion Funnel

Not getting a good night's sleep leaves us feeling fatigued the next day. If we suffer like this on a regular basis, we can end up driving ourselves down the Exhaustion Funnel, as illustrated by Marie Åsberg (see Figure 4.1), an expert on burnout from the Karolinska Institute in Stockholm. We can end up being irritable, or having unexplained physical symptoms, before sinking further into a feeling of joylessness or hopelessness. To stop ourselves being swallowed up by the black hole of exhaustion at the bottom of the funnel, there is something we can do. We need to take pre-emptive action by looking after ourselves more compassionately, with self-awareness being key in order to notice our moods and feelings and their impact on our thoughts and actions.

Åsberg suggests that we can nourish ourselves by choosing energising activities that we know will make a difference to our psychological wellbeing. Simply making a list of 'nourishing activities' in one column versus 'depleting ones' in another can help here. Then, if we feel ourselves slowly slipping down the path to exhaustion, we can engage in the more nourishing pursuits. What constitutes a nourishing activity will depend on the individual concerned: it could be listening to music, reading a book, or simply chatting to a friend. In parallel, we can seek to eliminate some of the more depleting activities. Examples could include watching TV passively for four hours on the trot or checking your smartphone 500 times a day. It is worth pointing out that activities like these may only be depleting if we overindulge in them: everything in moderation! Half an hour in front of your favourite soap may be the

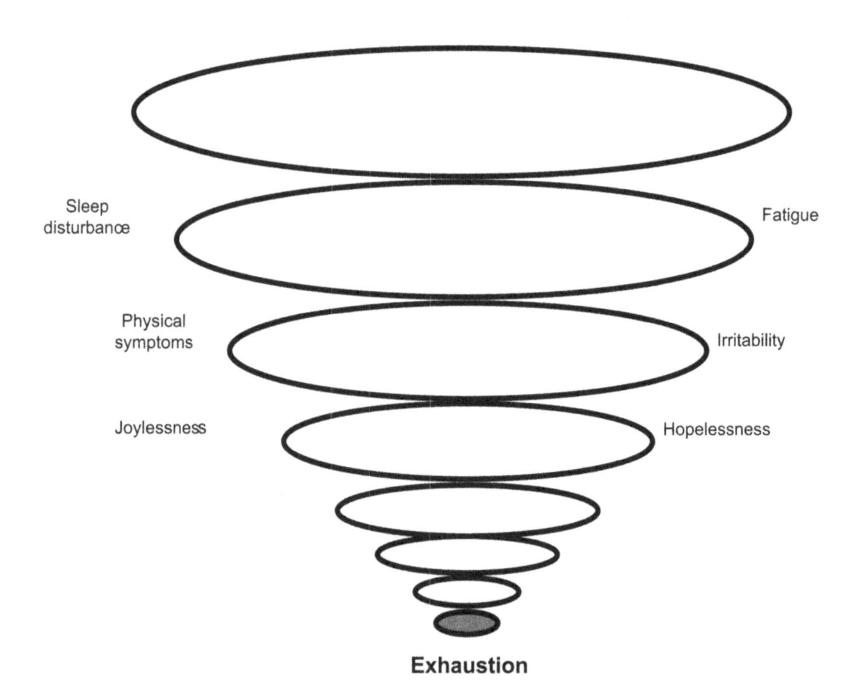

FIGURE 4.1 Åsberg's Exhaustion Funnel

perfect tonic, but sitting in front of the TV all night is a very different story, and you may well feel drained the next day at work, especially if your sleep has suffered.

You can see in the diagram that fatigue is one of the early warning signs and can set up a pathway to exhaustion. Since it is such a big topic in its own right, and an enemy of mindfulness, the next chapter is devoted to it.

Mental Health and Sleep

Many people will acknowledge the fact that too little time in the land of nod impacts their emotions and their ability to make sound decisions. But if we find ourselves on the slippery road to exhaustion when we are not sleeping properly, it can cause real mental health difficulties too. Sleep disruption has been found to precede depression. Disturbed sleep could in fact provide an early warning of mental health issues. In schizophrenics, sleep patterns can be taken to the point where they are totally smashed. There is some good news, however. Researchers at Oxford University have found that delusional paranoia can be reduced by 50 per cent if sleep is stabilised using cognitive behavioural therapy.[18]

Clocking Out

Resting involves 'clocking out'. In this state of mind, you are no longer on task and can give up being accountable to anyone for a while. The good news is that you may only need to nudge yourself into resting at opportune moments during the day. Your mind frequently needs to replenish itself, but this can be done in lots of ways. For example:

- When you wake up, remind yourself of your key purpose in life.
- Pause for 30 seconds after your breakfast and gather yourself before moving off again.
- Notice that breathing space between the end of an inhalation and the beginning of an exhalation.
- Give yourself a few moments of peace once you finish a task, before moving to the next one.
- Encourage your mind to rest by disengaging from 'chatter' about yourself or other people.
- Sit in silence for a minute each day.

There is a clear difference between rest and sleep, because sleeping is an unconscious activity and we have no choice over where our mind takes us. On the other hand, conscious effort is needed to rest purposefully, but this can pay dividends in re-energising ourselves for the challenges ahead.

DIGITAL HEALTH

This section takes a look at the potential use of digital technology in the business world and examines its pros and cons. Digital health is rapidly taking hold in the workplace, promising fitter and more productive employees. It can be an effective tool in helping us achieve a greater degree of mindfulness, but we may also find ourselves on a digital treadmill to mindlessness.

A healthy workforce could save companies money. For example, the pharmaceutical company McKesson concluded that its health and wellness programme saved almost $12 million in medical and productivity costs over two years.[19]

If digital technology is leveraged effectively, everyone can end up being a winner. We'll start by looking at the potential health gains.

Thinking Digitally

In our private lives, we are positively surrounded by digital technology, which enhances our working lives in a thousand ways via computers, smartphones and tablets. But modern workplaces often fail to fully harness the technology that could also significantly enhance our health and productivity at work. Bucking the trend, some large businesses have embraced the technology their employees already own. Bosses at computing giant IBM, the software company Autodesk and the retailer Target have encouraged their employees to track their steps with Fitbit fitness trackers.[20] If teams compete to rack up the number of daily steps they take, fitness and work productivity can improve in tandem.

Taking it a step further, some start-up companies are leveraging anonymised data on workers to help monitor health and make recommendations for employers. The 'Big Brother' element to what employers will end up knowing about our health sits uncomfortably with some people. Nevertheless, corporate health done digitally has the potential to be the catalyst for cultural change in the workplace. Digital technology can be used as a tool to promote mindfulness and help effect behavioural change too, amounting to what can be called a 'digital health mindset'.

Digital Corporate Health

The digital health mindset:

- Helps us monitor sleep and respects its importance.
- Exploits consumer technology, such as the Fitbit, in the workplace.
- Utilises the 'quantified self' as a tool for creating health benefits.
- Encourages healthy competition, with team incentives.
- Raises awareness of health metrics amongst employees.
- Improves health outcomes for employees cost effectively.

Fighting Inactivity

Employers could help fight workplace inactivity by encouraging their employees to move around more. According to the World Health Organisation (WHO), inactivity is the fourth biggest killer of adults.[21] The United Kingdom alone spends nearly £47 billion a year on obesity, which is more than on armed violence, war and terrorism.[22]

In fact, people who sit the longest may double their risk of diabetes or heart disease, as compared to those who sit the least.[23]

For those who sit for prolonged periods as an integral part of their jobs, such as drivers, it can be doubly difficult to achieve a satisfactory level of fitness.

The Quantified Self

Perhaps no one quite anticipated how much technology could reduce our lives to a set of numbers. Smartphones and activity trackers can measure all manner of things: steps taken, floors climbed, kilometres walked, time spent sitting down, heart rate and calories burned, to name a few. You'd be forgiven for thinking the quantified self is the whole self, and nothing but the self. It is not, of course, but the ability to track our every move enables effective goal setting and progress tracking. Electronic notifications and alerts can sometimes be a rude interruption in our working lives. However, when we are discreetly reminded to stretch our legs after a period of staring at our screens, the technology seems to have our best interests at heart. We can take a quick break and return to work feeling refreshed. Productivity goes up, as does the generation of creative ideas.

Activity trackers linking to smartphone apps are available without a subscription. Employers are often happy to subsidise gym memberships, but could this be a more effective, lower cost alternative? Non-attendance at the gym after the initial enthusiasm has worn off often becomes a source of guilt. We might be better off starting our exercise routines from the office: climbing the stairs, taking a brisk walk at lunchtime in a 'green gym' and walking to the station instead of taking the bus. It can be more motivational, too, to think in terms of micro-activities rather than workouts. Every micro-activity contributes something, and together they add up to a real health outcome. Largely unacknowledged micro-activities, such as cleaning your desk, taking a trip to the printer, or a walk to see a colleague, quickly add up to a significant calorie burning gain.

Corporate Calories

It is extremely rare for there to be corporate targets for calories burned in the course of a working day. But work is often where a genuinely competitive team spirit thrives. Calorie burning comparisons across individuals, teams and departments could usher in a healthier workplace, and would likely have a positive, knock-on effect on business performance too. The technology and the data are there, but they are not always leveraged for improved employee health. Blue-collar workers, who are typically more active than office-based employees, might enjoy their currently under-recognised status as high calorie burners. This thinking is backed up by science too.

In a fascinating study by researchers at Harvard University, a group of housekeepers were informed of how many calories they were burning during the day over a period of four weeks.[24] By simply telling them the value of their daily activity, they were actually able to lower their body fat, blood pressure and waist-to-hip ratio (as compared to a control group who were not given any information at all).

As unbelievable as it sounds, this 'placebo-style' effect actually created measurable, real-world gains. Apart from the information provided to the housekeepers, there was no other intervention. It is extraordinary to think that simply telling the housekeepers how many calories they had burned produced measurable, physical effects. It would make sense to use digital technology to leverage this interesting effect. There are plenty of devices out there that can count calories in order to exploit this finding. After all, just knowing the number of calories we are burning can lower our body fat!

HIJACKING YOUR BRAIN

All of our minds can be hijacked. Our choices are not as free as we think they are.

– Tristan Harris, ***design ethicist and former Google employee***[25]

We have explored some of the potential health gains from the technology in our lives. At the same time, there is a considerable 'flipside' to digital technology, which relates to 24/7 connectivity. We do not want to end up being digital automatons. Even the least observant amongst us will see people walking around mindlessly glued to their smartphone screens, if we are not bumping into them by mirroring their behaviour ourselves. Let's get a handle on how the technology is causing this mindlessness.

When was the last time you checked your smartphone? Fifteen minutes ago? It is usual now to become anxious when we are not mentally paired with our phones. That irresistible urge to check your phone has been likened to the effect of a slot machine. When you put your phone down, the adrenal gland in your brain releases cortisol. You then become anxious, just because of a signal sent by your brain. The chances are that you will respond quickly to avoid feelings of boredom, frustration or confusion. There is a tendency to attribute distractions to our surroundings, but the cravings we feel for our phones do not come from our outside environment. Our brain's reward centre wants its next dopamine hit to feel good again.

Once we have been hooked back in, Silicon Valley designers, software engineers and product managers know exactly what to do to maintain our brain's attention. Welcome to the 'attention economy'. The 'like' feature on Facebook only appeared in 2009, but it massively increased engagement. It was quickly copied by other social media platforms. The receptors in the brain grow accustomed to 'likes', scanning for the next post to provide the required surge of dopamine. Opioids, which attach to receptors to produce a morphine-like relaxation effect, also increase once your brain fixates on a target.[26]

Technology may be encouraging a form of addiction amongst users, and it may also be curtailing people's ability to focus. One study of 800 smartphone users showed that the mere presence of such a device could diminish cognitive capacity. Shockingly, this smartphone-induced 'brain drain' effect was still apparent when they were switched off.[27] If our cognitive capacities are being impaired, and our brains are being hijacked in this way, it is time to take action. The constant exposure to screens may well be damaging our health. The average American checks their phone 80 times a day, and one in ten check their phone every four minutes.[28]

It is time for a digital detox.

The Digital Detox

Learning to live with technology in a healthy way poses a considerable challenge. Breaking electronic habits can be hard, but not being mesmerised by a screen quite as much can make you feel quite different, providing the motivation to keep going.

Here are ten top tips for unshackling yourself:

1. Use a tracking app for your smartphone to measure how much time you spend looking at your screen, and how many times you pick it up. This will provide a baseline measurement before you make any changes.
2. Once you know how many hours your habit consumes, ask yourself what you might like to do with that time instead. The three weeks a year we typically spend checking social media could be used for a new hobby, or even a holiday.
3. Turn all banner-style, pop-up and sound notifications off on all your smartphone apps. If you wish, keep the badge-type notifications.
4. Delete social media apps from your phone. Use a desktop computer to check these instead.
5. On your commute, keep your phone hidden out of sight.
6. Do the same for meetings, conversations and meals out with friends or family.
7. Keep your phone out of sight when you are driving. Even hands-free technology can slow reaction times down, because people are dividing their attention.
8. Leave your gadgets at home occasionally to assist in freeing yourself from the constant checking. Use the time to pay more attention to your environment.
9. Implement a 'digital sunset'. Set a time each day, at least two hours before bedtime, for completely downing your electronic devices. Stick to it.
10. Store your devices outside your bedroom. This will stop you reaching for your screen in the middle of the night, or when you wake. Sleep issues can coexist with technology addiction.

You do not have to go cold turkey and implement these all at once – think about the incredible anxiety that could provoke! It may be best to break habits one or two at a time. Observe how the dropping of each one makes you feel. Constantly remind yourself that the likely benefits are improved mood, sleep and productivity, and check to see whether this is true for you.

USING OUR SENSES AGAIN

The ultimate aim of a digital detox is to reclaim some control over the way we allocate our mental resources. A detox can help us break free of all the stimuli clamouring for our attention. This is doubly difficult, since it is not just our devices that provide the stimuli. Our brain chemistry plays a significant role too. By becoming aware of our cravings and our moods, thoughts and feelings, we can assert some freedom once again. Then we can learn to be fully present in the bodies we inhabit. Mindfulness gives us that choice.

The One-Minute Mindfulness Meditation

Many people ask how they can start the practice of mindfulness. Dedicating just one minute to it can make a difference. You'll feel more able to take on the challenges of the day, whilst being able to work more safely.

Here is one way to get started, if you can find an opportune moment during the day:

- Sit upright in a straight-backed chair. It is best to allow your spine to be self-supporting by moving your back a short distance from the rear of the chair. Rest your feet flat on the floor. Close your eyes or, if you prefer, simply lower your gaze.
- Bring your attention to your breath as you inhale and exhale. What sensations are you feeling as you breathe in and out? Just observe what is happening without trying to label the experience in any way. Nothing special needs to happen. There is no need to modify your breathing in any way.
- If you find that your mind begins to wander, simply notice that this is happening, then bring your attention back to your breath. There is no need to be hard on yourself, or to criticise yourself in any way. Minds wander – that is just what they do. Noticing this is a central part of mindfulness meditation.
- Observe what your mind is doing. It might become calm like a still pond. But if it doesn't stay that way, resist the temptation to be critical of yourself. Whatever sense you have of what is going on for you, it may only be fleeting. If you feel angry or tired, simply observe this too. That too may just be fleeting. Whatever happens, just allow it to be.
- When a minute has passed, open your eyes. Slowly allow yourself to take in the room again. Return yourself to your surroundings.

KEY POINTS

- There are well-researched health benefits to practising mindfulness, and the evidence base from neuroscience is growing rapidly. The benefits cover the full range of physical and mental health conditions.
- Mindfulness is immediately applicable in everyday activities, although to experience the full benefits (and gain a colourful brain scan!), an eight-week course is recommended.
- Sleep is the cornerstone of good physical and mental health. You can sleep your way to better performance.
- Circadian rhythms are hard-wired into our evolutionary make-up; attempts to override them are unlikely to succeed.
- The Exhaustion Funnel is a useful way of observing the potential downward spiral from fatigue and a lack of sleep, to exhaustion and mental health problems.

- Digital health has many potential benefits. Metrics on the calories we burn, and the quality of our sleep, can help us monitor our progress towards becoming fitter and healthier. This could be further exploited in the workplace.
- There's also a dark side from over-engaging with electronic devices. They can hijack our brains. To counteract this, we may need to follow a digital detox to break unhelpful habits.
- Once we have digitally detoxed, we can reclaim the way we allocate our attentional resources through increased self-awareness.

M4: APPLYING MINDFUL SAFETY FOR BETTER SELF-CARE

Individual

- Practise everyday mindfulness with routine activities. Also try the one-minute mindfulness exercise as described in this chapter.
- Are you feeling sluggish? Notice how the amount of sleep you get affects your performance in everyday life and at work. Check in with yourself regularly.
- Notice if fatigue and sleep loss are beginning to contribute to a downward spiral. Are you experiencing a low mood? Can you prevent yourself from feeling any lower in good time?

Relational

- Consider buddying up with a friend with the goal of paying each other a little more attention than you currently do. Take a genuine interest in their life. Even asking them how they are by text can make a big difference.
- Feeling connected to others boosts our wellbeing, so make use of the support network around you. Provide the same support you would like to receive from others.

Organisational

- Does your organisation treat their staff with compassion? Investigate any initiatives they may have to help people who may be struggling to look after themselves effectively. Is the culture an understanding one?
- Exercise with colleagues. Go walking or running at lunchtime with them. If it works for you, use technology to set targets.

Societal

- Looking after ourselves also helps us attune our minds to the difficulties others may be facing, in both the home and the workplace, to society at large.
- Compassion for others and their struggles is part of mindfulness training, and it helps broaden our perspective.

NOTES

1. Kornfield, J. (1996). *Buddha's Little Instruction Book*, p. 28. London: Rider.
2. Davidson, R.J., Kabat-Zinn, J., Schumacher, J., Rosenkranz, M., Muller, D., Santorelli, S.F., Urbanowski, F., Harrington, A., Bonus, K. & Sheridian, J.F. (2003). Alterations in brain and immune functioning produced by mindfulness meditation. *Psychosomatic Medicine*, Volume 65, pp. 567–570.
3. Kabat-Zinn, J., Lipworth, L., Burncy, R. & Sellers, W. (1986). Four-year follow-up of a meditation-based programme for the self-regulation of chronic pain: treatment outcomes and compliance. *The Clinical Journal of Pain*, Volume 2(3), p. 159.
4. Speca, M., Carlson, L.E., Goodey, E. & Angen, M. (2000). A randomised, wait-list controlled trail: the effect of a mindfulness meditation-based stress reduction program on mood and symptoms of stress in cancer outpatients. *Psychosomatic Medicine*, Volume 62, pp. 613–622.
5. Bowen, S., Witkiewitz, K., Dilworth, T.M., Chawla, N., Simpson, T.L., Ostafin, B.D., Larimer, M.E., Blume, A.W., Parks, G.A., & Marlatt, G.A. (2006). Mindfulness meditation and substance use in an incarcerated population. *Psychology of Addictive Behaviours*, Volume 20, pp. 343–347.
6. Baer, R.A., Smith, G.T., Hopkins, J., Kreitemeyer, J. & Toney, L. (2006). Using self-report assessment methods to explore facets of mindfulness. *Assessment*, Volume 13, pp. 27–45.
7. Shapiro, S.L., Oman, D., Thoresen, C.E., Plante, T.G. & Flinders, T. (2008). Cultivating mindfulness: effects on well-being. *Journal of Clinical Psychology*, Volume 64(7), pp. 840–862.
8. Hick, S.F., Segal, Z.V & Bien, T. (2008). *Mindfulness and the Therapeutic Relationship.* New York: Guildford Press.
9. Hölzel, B.R., Carmody, J., Vangel, M., Congleton, C., Yerramsetti, S.M., Gard, T., Lazar, S.W. (2010). Mindfulness leads to increases in regional brain gray matter density. *Psychiatry Research: Neuroimaging*. doi:10.1016/j.pscychresns.2010.08.006. Epub 2010 November 10.
10. The Sleep Council (2013). The Great British bedtime report. Available from https://www.sleepcouncil.org.uk/wp-content/uploads/2013/02/The-Great-British-Bedtime-Report.pdf. (Accessed 13.01.2019).
11. Sulleyman, A. (2017). Netflix's biggest competition is sleep, says CEO Reed Hastings. 19 April 2017. *The Independent*. Available from https://www.independent.co.uk/life-style/gadgets-and-tech/news/netflix-downloads-sleep-biggest-competition-video-streaming-ceo-reed-hastings-amazon-prime-sky-go-a7690561.html (Accessed 18 March 2018).
12. Moller-Levet, C.S., Archer, S.N., Bucca, G., Laing, E.E., Slak, A., Kabiljo, R., Lo, J.C.Y., Santhi, N., von Schantz, M., Smith, C.P. & Dijk, D.J. (2013). Effects of insufficient sleep on circadian rhythmicity and expression amplitude of the human blood transcriptome. *Proceedings of the National Academy of Sciences*, Volume 110, pp. E1132–E1141.
13. Kroker, K. (2007). *The Sleep of Others and the Transformations of Sleep Research.* Toronto: University of Toronto Press.
14. Ericsson, A., Krampe, R.T. & Clemens, T.R. (1993). The role of deliberate practice in the acquisition of expert performance. *Psychological Review*, Volume 100(3), pp. 363–406.
15. Gallagher, J. (2017). Body clock scientists win Nobel prize. *BBC News Health*. Available from http://www.bbc.co.uk/news/health-41468229 (Accessed 10.02.2018).
16. Gallagher, J. (2017). Heart surgery survival chances 'better in the afternoon'. *BBC News Health*. Available from http://www.bbc.co.uk/news/health-41763958 (Accessed:10.02.018).
17. Gallagher, J. (2017). Daytime wounds 'heal more quickly'. *BBC News Health*. Available from http://www.bbc.co.uk/news/health-41918368 (Accessed 10.02.2018).

18. De Lange, C. (2016). Sleep well – your mind could do with it. *The New Scientist.* 28 May 2016, p. 39.
19. Rutkin, A. (2016). Fitter, more productive? *The New Scientist*, 28 May 2016, pp. 16–17.
20. Ibid.
21. Dugan, E. (2013). One small step…Inactivity is the world's fourth biggest killer. *The Independent*, 4 May 2013.
22. Burgess, K. (2014). £47bn cost of obesity is a bigger economic burden than terror. *The Times*, 21 November 2014.
23. Sitting for long periods 'is bad for your health'. *BBC News Health.* 15 October 2012. Available from http://www.bbc.co.uk/news/health-19910888 (Accessed 14.02.2018).
24. Crum, A.J. & Langer, E.J. (2007). Mind-set matters: exercise and the placebo effect. *Psychological Science*, Volume 18(2), pp. 165–171.
25. Lewis, P. (2017). Our minds can be hijacked: the tech insiders who hear a smartphone dystopia. *The Guardian.* 6 October 2017. Available from https://www.theguardian.com/technology/2017/oct/05/smartphone-addiction-silicon-valley-dystopia (Accessed 04.08.2018).
26. Pillay, S. (2017). How to win the smartphone-brain battle. *Psychology Today*, 13 July 2017. Available from https://www.psychologytoday.com/blog/debunking-myths-the-mind/201707/how-win-the-smartphone-brain-battle (Accessed 14.02.2018).
27. Ward, A.F., Duke, K., Gneezy, A. & Bos, M.W. (2017). Brain drain: the mere presence of one's own smartphone reduces available cognitive capacity. *Journal of the Association for Consumer Research*, Volume 2(2), pp. 140–154. doi:10.1086/691462.
28. Americans check their phones 80 times a day: study. *New York Post.* 8 November 2017. Available from: https://nypost.com/2017/11/08/americans-check-their-phones-80-times-a-day-study/ (Accessed 15.02.2018).

5 Fatigue
Safety's Silent Saboteur

Sleep deprivation is no joke.

– Áine Cain[1]

As we saw in the last chapter, sleep and fatigue are intimately connected. Particularly in high-hazard industries, it is imperative that we overcome fatigue's effects in order to prevent disasters from happening. It has played a significant role in major accidents involving the nuclear industry, the rail industry and space shuttle launches. We'll also look at some practical remedies whilst outlining the role of fatigue management to avoid the truly staggering costs of getting it wrong. Three main levels of the M4 approach are covered here: individual, organisational and societal. Societally, it is not an issue we can afford to ignore, and greater public understanding would significantly help in this regard.

THE MYTH OF SLEEP DEPRIVATION

Try going to sleep before reading any more of this chapter.

Still here? Unless you are watching the pendulum of a gold watch swing before your eyes, it is highly unlikely you can just fall asleep 'at will'. Very few individuals could manage this feat using willpower alone. Herein lies one of the problems of getting high-quality sleep, with the intention of arriving at work the next day as fresh as a daisy – you can't just order 'sleep to go'. Fatigue can be defined as 'extreme tiredness that affects one's ability to concentrate and work effectively'. It is an enemy of mindfulness for a reason. Our minds can't function reliably under its influence, and this even affects the performance of simple tasks. Having said that, we will each experience its effects in slightly different ways, both physically and psychologically.

Many world leaders have earned their hardworking reputations, at least in part, from having as little as four hours' sleep a night. This elite club of sleep-deprived individuals includes President Donald Trump and Margaret Thatcher. In the business world, similar stories abound, with the CEOs of AOL, PepsiCo, Fiat and the founder of Twitter all burning the candle at both ends – and apparently achieving great things. Virgin Group Founder Richard Branson is said to need only five or 6 hours a night, but this compares positively with the others.[2]

But a closer look reveals 'sleep debt' may be less than a perfect advert for high-flying endurance. Poor decision-making and errors of judgement are the likely consequences. We may occasionally like to boast chirpily how little sleep we have had, but our bloodshot eyes often betray our implied mental alertness. Being awake for 17 hours can impair our performance to the same degree as two units of alcohol, or a pint of lager.[3]

Impaired Performance

The general consensus is that individuals need an average of seven to eight hours' sleep per night. If this is reduced by one or two hours, the typical result is a deterioration in alertness and performance the next day. If sleep deprivation continues in the longer term, it can contribute to poor physical and mental health, as well as ongoing difficulty concentrating on the job in hand. Research by the US military has shown how productivity on a routine mechanical task falls rapidly for people restricted to less than seven hours' sleep a night. Over a period of 20 days, the group surviving on seven hours found their productivity fell to around 85 per cent of their first day's level.[4] The group surviving on just four hours' slumber could only manage around 15 per cent of their original productivity level. This is an alarming drop, especially considering that the routine task did not involve complex decision-making or higher cognitive functioning. As we shall see later on in this chapter, you certainly do not want to be launching space shuttles on such little sleep.

A Biological Need

Biologically, we have evolved a circadian rhythm, which affects our physiological processes over the course of a day.[5] In practice, this means that mental alertness in fully rested individuals tends to be lowest around 4 am. Our biological need for sleep is greatest in the early hours of the morning, but we also experience a noticeable drop in alertness mid-afternoon. Even if you have never done shift work, you are likely to have experienced the same effects on your body from a long-haul flight – in other words, jet lag. Disrupted sleep and hunger patterns are the usual consequences as the body attempts to adjust itself to a new routine.

Our body clock is largely governed by daylight, and this is why it is far more difficult to regulate than a regular clock. We can't reprogram our bodies like we can a clock. Fly to a different time zone or change your shift pattern, and the body clock stubbornly resists change as the brain plays catch-up.

STRATEGIC NAPS

In Mediterranean cultures, siestas are not just cultural adaptations to a hotter climate, but a way for the body to replenish its energy supplies. We might all benefit from a 'siesta style' afternoon nap, especially if we are suffering from a lack of sleep, but whether you call it a siesta or a strategic nap, it could make a huge difference to the safe, competent performance of your work. We know that shift-workers are more likely to build up a sleep debt than nine-to-five workers.[6] In fact, research shows that shift-workers who sleep in the daytime will experience lower quality sleep, typically sleeping for a third less than they would at night.[7] Moreover, they tend to wake up spontaneously after fewer hours' sleep.

In these circumstances, strategic napping can play a hugely positive role in restoring energy. Even after 40 hours of sleep deprivation, a two-hour nap can maintain performance at 70 per cent of well-rested levels.[8] Please note that I am not in any way suggesting you deprive yourself of sleep for 40 hours then nap for two hours to test this out!

Tips for Exploiting the Strategic Nap

- Create the right environment
- Time your naps appropriately
- Take your naps early on
- The longer the better
- Choose between 45-minute naps and two-hour ones[9]

Create the right environment. This may be easier to arrange at home than it is in operational settings. You are going to need somewhere that is dark and comfortable. If lighting and noise are an issue, sleep masks or foam earplugs may be able to control the effects. Keeping distractions that may disrupt napping quality to a minimum is key to a good experience.

Time your naps appropriately. Think in terms of circadian rhythms. Night-time naps are best taken between 1 am and 6 am. Daytime naps should ideally be taken between 2 pm and 4 pm. In both cases, these times coincide with natural dips in alertness, making a sleep-like quality easier to maintain.

Take your naps early on. The longer you are awake, the more the effects of fatigue will impact you. If you rise at 6 am and start a shift at 10 pm, you will have been awake for 16 hours by the time the shift starts. If your shift is eight-hours long, you will then be awake for 24 hours. In this case, it would be smart to take a two-hour nap at 2 pm, as it will harmonise better with your circadian rhythm. And by the time you start your shift, you will have been awake for only 6 hours.

The longer the better. Just as for sleep, the longer the better. Seven hours will trump four hours every time. And a 30-minute nap will be far more effective than a 10-minute one. For longer naps, see the advice below.

Choose between 45-minute and two-hour naps. After 45 minutes, the average person will fall from light into deep sleep, which is much more difficult to wake up from. That post-nap grogginess (sleep inertia) can be avoided if a 45-minute nap is cut off at this point. After 100 minutes, most people will be cycling out of the deep sleep phase, so waking up 20 minutes later at the end of a 2-hour nap is easier.

FATIGUE KILLS

There's no blood test for it and you can't be breathalysed for it, but the effects of fatigue on our ability to remain alert and do our jobs properly compare with those of alcohol. Fatigue has also been cited as a big contributor to major accidents, so educating people about the potential consequences should be on everyone's agenda. As we shall see, fatigue can influence life or death decisions made at nuclear power stations, at space shuttle launches and on the railways.

Chernobyl

Fatigue can sometimes be an elusive factor to pin down, but it is implicated in the world's worst nuclear accident at Chernobyl in 1986. The catastrophe began at 1:23 am on 26 April, as the result of human error, which led to one of the reactors being placed in a volatile state.[10] In addition, many organisational factors played a role, but we are interested here in how fatigued staff may have performed mindlessly at that time. Remember that in the small hours of the morning, our biological need for sleep is greatest.

The mental alertness of the Chernobyl operators would naturally have been much lower than in normal waking hours. It may be difficult to draw any firm conclusions about the contribution of fatigue-related errors in this case, but such mistakes are far more likely to occur at night.

Space Shuttles: Challenger and Columbia

To send a shuttle into space, you need a good night's sleep. Fatigue played a significant role in the Challenger space shuttle disaster on 28 January 1986. The shuttle broke up just 73 seconds after launch, killing all seven crew members. Warnings about launching the shuttle in the low temperature that morning were ignored. The immediate cause was the failure of the O-ring seal, which allowed pressurised burning gas to escape, but some of the key managers involved in the launch had fewer than two hours' sleep the night before. The Presidential Commission's report cited the contribution of human error and poor judgement related to sleep loss and shift work during the early morning hours prior to launch.[11]

The role of fatigue was also highlighted in the near-catastrophic launch of the Columbia space shuttle just days before the Challenger disaster.[12] On that occasion, no less than 18,000 pounds of liquid oxygen were accidentally drained from the shuttle's external tank a few minutes before launch. Disaster was narrowly averted, but the liquid oxygen loss went undetected until just 31 seconds before lift-off. It was the console operators' third day of working 12-hour night shifts, and they had been on duty 11 hours at the time of the critical error. Long, consecutive night shifts produce an environment where mistakes are easily made. Sadly, lessons could not be learned in time for the Challenger's fatal mission.

Clapham Junction

In Britain, the Clapham Junction Rail crash in 1988 was responsible for 35 people losing their lives and 500 being injured. A signal failure caused by a wiring fault might not immediately spring to mind as fatigue related, but the signalling technician responsible had worked a seven-day week for the previous 13 weeks. In all that time he'd had only one day off. One day off in 13 weeks! Many other failings at a supervisory and management level were uncovered by the official report into the crash, but the technician's fatigue was cited as a significant factor. The constant repetition of weekend work was said to have, "Blunted his working edge, freshness and concentration".[13]

SHIFT-WORK AND SLEEPINESS

As we have seen, sleep loss and fatigue have played a significant role in major accidents. Whether it was overseeing a nuclear facility, launching a space shuttle, or wiring up a railway, the shift-work was unavoidable for 24/7 operations. The research findings are unequivocal: shift-work causes sleep disturbances and fatigue.[14–17]

Shift-workers often experience difficulties in going to sleep and then maintaining and consolidating it.[18] Unsurprisingly perhaps, up to 90 per cent of them report sleepiness at work, whilst 20 per cent report involuntarily falling asleep there too.[19] Despite the pitfalls that go with the territory, there is something these workers can do to align themselves to shifts that work best for their personalities.

Are You a Morning Lark or Night Owl?

Morning larks are more alert and energetic in the morning, whilst night owls will only peak later on in the evening. Matching yourself up to the right shifts could make a huge difference to the length and quality of your sleep, ultimately leading to greater health and wellbeing. One research study of 238 shift workers found a significant relationship between these two types of people and the social jet lag, sleep duration and sleep quality experienced.[20] As expected, night owls experienced greater social jet lag, shorter sleep duration and poorer sleep quality on morning shift days. A similar pattern was found for morning larks working on night-shift days.

There are important implications here for high-hazard industries like the ones discussed earlier in this chapter. Sleepiness and fatigue are major risk factors for workplace injuries and accidents. They lead to impaired cognitive performance on a range of measures, such as decision-making, judgement, reasoning, vigilance, memory, learning and motor skills.[21,22] To ensure that shift-workers get a good night's sleep, workplaces must intelligently match shifts appropriately to the needs of morning larks and night owls.

THE PURSUIT OF DAYLIGHT

Many people complain about their morning grogginess in the week following the spring clock change. It is our sleep that tends to suffer in the short term, as we play catch up after the loss of an hour. In the autumn, we trade our longer evenings for brighter mornings. At first sight, it may seem like a fairly innocuous subject. In fact, Daylight Saving Time (DST) has been hotly debated for over a century, and it is a fascinating, controversial subject that engages the politics around agriculture, energy and health and safety.

Back in the 1930s, even Winston Churchill was once tempted to wade into the debate, stating that DST could enlarge "the opportunities for the pursuit of health and happiness among the millions of people who live in this country".[23] He was highlighting the benefit of that extra daylight in the evenings for the pursuit of leisure. But over the years, not everyone has been such a staunch advocate of making the change. Dairy farmers, for example, have tended to oppose DST, as it affects their carefully arranged milking schedules. Look at it from a dairy cow's perspective. You are used to being milked at 5 am, and when the clocks go back in the autumn, you have to wait an extra hour. Naturally, you are positively bursting. Just as you get used to the new schedule, it changes again in the spring. This may all seem like a humorous digression (not if you are a cow!), but what it does is illustrate the extent to which DST affects the whole of society.

Road Safety

A comprehensive New Zealand study looked at over 12 million accident claims made during 2005–2016.[24] Road accident rates were up a significant 16 per cent on the first day of DST, and up 12 per cent on the second day. These results make intuitive sense. After losing a chunk of sleep when the clocks go forward, we are groggy and less mentally alert, and this is likely to cause a higher number of road accidents. However, the picture is more complex than that because we need to factor in the long-term effects too.

Long-Term Effects

After drivers have adjusted to the spring clock change – and this may take a week or two – the shift to brighter evenings may start to pay dividends. One systematic review of 24 studies (drawn mostly from the United States) suggests the long-term effects on road safety are positive.[25] An extra hour of daylight in the evening could lead to an overall net reduction in accidents. We know that driver performance deteriorates in poor light and reaction times are slower. Pedestrian activity is also higher in the evenings. More daylight increases the likelihood of spotting a pedestrian in good time, potentially saving lives.

One US study suggests that year-round DST could reduce overall pedestrian fatalities by 13 per cent.[26] And according to an authoritative UK study, year-round DST could also reduce overall fatalities by around three per cent.[27] These arguments have been soundly made before, but DST policy recommendations must take account of more than just the issue of road safety, as we have already noted.

NOTICING FATIGUE AND PREVENTING BURNOUT

We have discussed how the effects of fatigue are similar to those experienced under the influence of alcohol. Lower standards of performance become acceptable, and we become increasingly incompetent on the job, though we are less conscious of this fact. We therefore need to intensively train our minds to notice when our performance is dipping below an acceptable level. What do we need to look out for?

When we are tired our attention wanes, as does our ability to maintain situational awareness. Logical reasoning is impaired. Attitude and mood deteriorate – we get grumpy! Tasks become more difficult to perform in a timely and accurate way, and we run the risk of involuntary lapses into sleep. Many of us will have experienced this on a long drive home late at night or in the early hours of the morning.

Because our ability to judge our own performance deteriorates as we become even more fatigued, we have to rely on recognising the symptoms as early as possible. We also need to evaluate our habits and consider changing them.

How to Prevent Burnout

If you find yourself regularly experiencing physical symptoms, such as drowsiness and your eyes closing involuntarily, it is time to act. The slope towards possible burnout is a slippery one. Long-term health is at stake, and lives may be being put at risk too.

- Get enough sleep
- Set up a bedroom routine
- Get enough exercise
- Eat and drink well
- Increase social contact

Get enough sleep. Aim for eight hours. Some individuals will need slightly less and some slightly more. Remember that just an hour less than your usual requirement can start to impact your health at the genetic level after only a week. Blood test research reveals that around 500 genes are switched on or off by changes in sleep patterns. A good night's sleep is enormously important. You can't cheat your body and mind if you are sleep deprived.

Set up a bedroom routine. Bedrooms are for sleeping, or so you'd think – not so much in many households these days. Large-screen TVs and an invasion of smartphones and tablets can make bedrooms anything but restful. These devices are also known for emitting blue light, which interfere with sleep. Enforce the discipline of a 'digital sunset' to restore a good night's sleep to the bedroom. In addition, there's no harm in setting the alarm clock to remind you to start your bedroom winding down routine each night. Ensure you block out any light, using blackout blinds if necessary, to avoid interfering with the body's circadian rhythm. Take steps to block out any disturbing noise too.

Get enough exercise. Exercise plays an important role in regulating our sleep.[28] Just avoid doing any within two to three hours of your bedtime. Early morning exercise is a different kettle of fish, though. Exercising before work starts can increase alertness on the job. Work out the best way of integrating activity into your routine. Cycling to work or walking there briskly as part of your commute can boost your health without you having to change your routine too much. There has also been a proliferation of smartphone workout apps that make routines less daunting. A personal favourite of mine is a seven-minute workout that makes it far easier to find the discipline to get started.

Eat and drink well. A balance of healthy foods and plenty of water can stop our energy levels from fluctuating too much. Foods with high sugar content and refined carbohydrates provide an initial boost, but they do not deliver sustained energy over a longer period.[29] Dehydration lowers our level of alertness. An alcoholic drink or two before bedtime might seem like a good idea in order to relax, but sleep quality is affected. Eating and drinking mindfully means being alert to what the body needs. Fully appreciate what you consume so you do not overdo it. Stay away from the vending machine!

Increase social contact. If you find your attention flagging, another great tip to fight on-the-job tiredness is to start a conversation with co-workers. The social interaction is likely to provide a much needed boost if you are suffering from mid-afternoon drowsiness.[30] Due to the body's circadian rhythm, we tend to experience a drop in alertness at this time.

Google's Example

Google sets itself apart from conventional companies in many diverse ways. It is famous for encouraging dogs in the workplace, 'all-hands meetings' and unscripted Q&As with executives. There's also an atmosphere of fun and creativity.[31] But Google's workplace practices also have a clear role to play in preventing fatigue and burnout. Lessons from Google can be applied in other industries well beyond Silicon Valley's technology hotbed.

Take a look at the long list of freebies Google provides: gyms, free meals, haircuts, dry cleaning and car washes on campus. By looking after their employees' errands in this way, a powerful message is being sent: this is all arranged so you can focus better on your work. If time is precious, carrying out those fatigue-inducing, mundane chores outside work can make stress levels rise. Google's approach is designed to create a better work-life balance. They once operated a novel kind of workplace flexibility: the 20 per cent rule. This allowed employees to take a day off a week to pursue personal projects. Though this isn't the practice anymore, it has metamorphosed into learning about other Google projects or departments.[32] Employees remain mentally stimulated in this way, providing an effective countermeasure against burnout.

Workplaces that are prescriptive in telling employees how to work will pay the price in lost creativity. Google allows people to work according to their unique needs and personalities. They can work from almost anywhere. That could mean working from a café, a beanbag seat, or even a swing. Creating such conditions for 'ease of being' is also part of mindfulness practice. Authentic selves are more likely to thrive at work, and fatigue is far less likely to make an inroad into the daily grind.

RELAXING THE MIND

Modern life's 'sleep arrogance' is finally being challenged. It can be taken one step further, however. Learning to rest properly and relax our minds may be almost as important.

> Our minds must relax: they will rise better and keener after a rest. Just as you must not force fertile farmland, as uninterrupted productivity will soon exhaust it, so constant effort will sap our mental vigour, while a short period of rest and relaxation will restore our powers. Unremitting effort leads to a kind of mental dullness and lethargy.

The quote above was not written by the latest health and wellbeing expert, but by a Stoic philosopher called Seneca, who lived around 2,000 years ago.

Recharging Your Batteries

It can be hard to find the time to recharge our batteries. Here's how to approach the subject:

- Accept the need to rest
- Stop and shed your 'to do' list
- Forget about the future
- Challenge your beliefs about rest.

Accept the need to rest. It might be glaringly obvious, but we all need to rest. And I am not talking about getting enough sleep here. In some ways, sleep is the easy way out – just hit the pillow when you are exhausted! By that stage, the only alternative may be to prop your eyelids open with matchsticks.

Stop and shed your 'to do' list. I mean stop completely, whilst fully awake, completely forgetting that shedload of 'must-dos'. Not many of us have time to properly rest these days, what with all the competing demands for our attention. If I'm honest, it is probably been a while since I achieved that restful state of mind. But failing to rest often enough can leave us feeling run down and worn out. It may even make us ill.

Forget about the future. When was the last time you totally forgot about planning, going somewhere or achieving something? That restful state of mind is hard to achieve in practice. We want to attain that sense of just being, without the 'thought baggage' of things not being completed on time. But to become more resilient, to toughen up for the challenges ahead, we need to rest. Rest can put us back on the road to achieving our goals, however counterintuitive this may sound. Our personal bucket lists can always wait. Paradoxically, it may take some work to rest.

Challenge your beliefs about rest. We all carry our own beliefs about having a rest. These may include beliefs that others will judge you, you'll let somebody down, or that you'll fail to keep up with everything. Underpinning those beliefs, there may be a central belief that it simply isn't OK to rest – this needs challenging. Give yourself permission.

KEY POINTS

- Fatigue can silently sabotage our efforts to perform safely. Thinking, judgement and concentration are all likely to become more clouded when we are sleepy. We cannot be mindfully present in these conditions.
- Not sleeping enough has been over-hyped and is often associated with boasts of being able to get more done. Macho attitudes in society are highly questionable.
- Barring a few exceptions, the reality is that sleep loss and fatigue severely affects individual performance.
- There have been several major accidents where fatigue has played a significant role. Organisations must take this on board to provide effective fatigue management.
- Road safety is clearly affected by fatigue and poor lighting conditions.
- To prevent fatigue and possible burnout, it may be necessary to change some of our lifestyle habits.

- Workplace environments can foster creativity and engagement, making fatigue less likely. Google's famed approach could be applied in other settings.
- We can also recharge our batteries by scheduling rest.

M4: APPLYING MINDFUL SAFETY TO BEAT FATIGUE

Individual

- Notice how fatigue affects your concentration, thinking and judgement.
- As a rule, get plenty of sleep – for most people, this will be around eight hours.
- Learn to nap strategically if you need to.

Relational

- Ask your partner, friend or colleague to provide some feedback on your current state of mind and physical condition, especially if you are doing something safety critical. They may spot symptoms of fatigue, even when you do not.
- Observe symptoms of fatigue, such as bloodshot eyes and clumsy movements, in others. Talk to them about it. Take an interest in the amount of sleep they have had.

Organisational

- Does your organisation have good fatigue management policies? Find out more and help improve things if there are gaps.
- Are any shift patterns intelligently devised for morning larks and night owls?

Societal

- Does your organisation provide education on the lifestyle factors that may affect sleep quality and fatigue?
- Do they provide napping facilities or places to rest properly?
- Is flexible working allowed to help staff get the rest they need and engage with their work more?
- Challenge societal myths about surviving on only a few hours' sleep.
- Ask what attitudes prevail about sleep and fatigue in your culture.

NOTES

1. Cain, Á. (2017). 11 successful people who hardly get by on any sleep. *Business Insider*. 26 July 2017. Available from http://uk.businessinsider.com/successful-people-who-do-not-sleep-2017-7/ (Accessed 18.02. 2018)
2. Ibid.

3. Dawson, D. & Reid, K. (17 July 1997). Fatigue, alcohol and performance impairment. *Nature*, Volume 388, p. 235.
4. Belenky, G. (1997). Sustaining performance during continuous operations: The US army's sleep management system. *Managing Fatigue in Transportation, International Conference Proceedings*, pp. 95–103.
5. Moller-Levet, C.S., Archer, S.N., Bucca, G., Laing, E.E., Slak, A., Kabiljo, R., Lo, J.C.Y., Santhi, N., von Schantz, M., Smith, C.P. & Dijk, D.J. (2013). Effects of insufficient sleep on circadian rhythmicity and expression amplitude of the human blood transcriptome. *Proceedings of the National Academy of Sciences*, pp. E1132–E1141.
6. Rail Safety Standards Board (2014). T059 Human factors study of fatigue and shift work. Guidelines for the management and reduction of fatigue in train drivers. *RSSB research report.*
7. Rail Safety Standards Board (2014). T699 Shift patterns of freight & infrastructure workers: Implications for fatigue & safety. *RSSB research report.*
8. Angus, R.G., Pigeau, R.A. & Heslegrave, R.J. (1992). Sustained-operations studies. From the field to the laboratory. In Stampi, C. (Ed.), *Why We Nap*, pp. 217–241. Boston, MA: Kirkäuser.
9. Caldwell, J.A. & Caldwell, J.L. (2003). *Fatigue in Aviation. A Guide to Staying Awake at the Stick.* Aldershot: Ashgate.
10. INSAG-7 (1992). *The Chernobyl Accident, Updating of INSAG-1: A Report by the International Nuclear Safety Advisory Group.* Available from https://www-pub.iaea.org/MTCD/publications/PDF/Pub913e_web.pdf (Accessed 24.10.2020).
11. Rogers Commission (6 June 1986). *Report of the Presidential Commission on the Space Shuttle Challenger Accident.* Available from https://spaceflight.nasa.gov/outreach/SignificantIncidents/assets/rogers_commission_report.pdf (Accessed 05.04. 2018).
12. Ibid.
13. Hidden, A. (1989). *Investigation of the Clapham Junction Railway Accident. Department of Transport.* ISBN 978-0-10-108202-0. Available from https://www.weathercharts.org/railway/Clapham_Junction_Collision_1988.pdf (Accessed 05.04.2018).
14. Carskadon, M.A. & Dement, W.C. (1981). Cumulative effects of sleep restriction on daytime sleepiness. *Psychophysiology*, Volume 18, pp. 107–113.
15. Carskadon, M.A. & Dement, W.C. (1987). Daytime sleepiness: quantification of a behavioral state. *Neuroscience and Biobehavioural Review*, Volume 11, pp. 307–317.
16. Czeisler, C.A., Allan, J.S., Strogatz, S.H., Ronda., J.M., Sánchez, R., Ríos, C.D., Freitag, W.O., Richardson, G.S. & Kronauer, R.E. (1986). Bright light resets the human circadian pace-maker independent of the timing of the sleep-wake cycle. *Science*, Volume 233, pp. 667–671.
17. Czeisler, C.A., Johnson, M.P., Duffy, J.F, Brown E.N, Ronda, J.M. & Kronauer, R.E. (1990). Exposure to bright light to treat physiological maladaptation to nightwork. *New England Journal of Medicine*, Volume 322(18), pp. 1253–1259.
18. Åkerstedt, T. (1998). Shift work and disturbed sleep/wakefulness. *Sleep Medical Review*, Volume 2, pp. 117–128.
19. Åkerstedt, T. (1995). Work hours and sleepiness. *Clinical Neurophysiology*, Volume 25, pp. 367–375.
20. Juda, M., Vetter, C. & Roenneberg, T. (April 2013). Chronotype modulates sleep duration, sleep quality, and social jet lag in shift-workers. *Journal of Biological Evidence*, Volume 28(2), pp. 141–151. doi:10.1177/0748730412475042.
21. Berger, A. & Hobbs, B. (2006). Impact of shift work on the health and safety of nurses and patients. *Clinical Journal of Oncology Nursing*, Volume 10, pp. 465–471.
22. Folkard, S. & Tucker, P. (2003). Shift work, safety and productivity. *Occupational Medicine*, Volume 53, pp. 95–101.
23. Churchill, W.S. (1934). A silent toast to William Willett. *Pictorial Weekly*. 28 April.

24. Robbs, D. & Barnes, T. (2017). Accident rates and the impact of daylight saving time transitions. *Accident & Prevention*, Volume 111, February 2018, pp. 193–201.
25. Carey, R.N. & Sarma, K.M. (2017). Impact of daylight saving time on road traffic collision risk: a systematic review. *BMJ Open*. doi:10.1136/bmjopen-2016-014319.
26. Coate, D. & Markowitz, S. (June 2004). The effects of daylight and daylight saving time on US pedestrian fatalities and motor vehicle occupant fatalities. *Accident Analysis & Prevention*, Volume 36, pp. 351–357.
27. Broughton, J. & Stone, M. (1998). *A new assessment of the likely effects on road accidents of adopting SDST.* Transport Research Laboratory. Available from https://trl.co.uk/reports/TRL368 (Accessed 11.12.2018).
28. Breus, M.J. (2013). Better sleep found by exercising on a regular basis. *Psychology Today*. Available from https://www.psychologytoday.com/blog/sleep-newzzz/201309/better-sleep-found-exercising-regular-basis-0 (Accessed: 17 February 2018).
29. Fisher, R. (2013). How to eat for more energy. *BBC Goodfood*. Available from https://www.bbcgoodfood.com/howto/guide/how-eat-more-energy-0 (Accessed: 17.02.2018).
30. Health and Safety Executive (n.d.). *Hints and tips for shift-workers*. Available from http://www.hse.gov.uk/humanfactors/topics/shift-workers.htm (Accessed 17.01.2018).
31. Wickre, K. (2017). What Google's open communication culture is really like. *Wired*. 20 August 2017. Available from https://www.wired.com/story/what-googles-open-communication-culture-is-really-like/ (Accessed 18.02.2018).
32. Wright, C. (2017). 10 Things to know about Google's awesome culture. *The Huffington Post*. 2 May 2017. Available from https://www.huffingtonpost.com/entry/10-things-to-know-about-googles-awesome-culture_us_59088802e4b03b105b44bbfd (Accessed 18.02.2018).

6 Distracted Minds, Lost Lives

It is clear that attention has become an acute collective problem of modern life – a cultural problem.

– Matthew Crawford[1]

I will start this chapter by sharing the example of what happened when a high-speed train was derailed with a distracted driver at the controls. We'll begin by focusing on the individual level, but it would be short-sighted to exclude the organisational context surrounding the train driver's lapse in concentration and loss of situational awareness. And the truth in modern life is that we are often prey to distractions from electronic devices perfectly designed to interrupt our attention – this is potentially life threatening when we are driving or must attend to any other safety-critical task.

The good news is that our minds can be trained to direct our attention and become far less susceptible to everything else clamouring for a say in where we chose to focus our energies. Mindfulness is an enjoyable practice in its own right, but it is also a defence against a multimedia onslaught that is responsible for creating a flow of irresistible distractions. It is a way of reclaiming what we choose to pay attention to, and tailored training interventions suggest it is highly effective in high-hazard, safety-critical environments where safe decisions must be made. One landmark study of train drivers in Spain is used to highlight the possible safety gains made as a result of enhanced focus and attention.

In relation to driving a vehicle, there is some promising research showing how greater empathy and shifting perspectives – qualities that mindfulness seeks to enhance – can positively impact safety. To experience the benefits, we must guard against a dependence on smartphone technology and approach self-driving cars with a degree of caution. To be fully present, technology will have to take a backseat.

Case Study: The Santiago de Compostela High-Speed Derailment

On 24 July 2013, a high-speed train travelling at around 190 kmh (118 mph), on its way from Madrid to Ferrol, derailed on a sharp curve three miles from Santiago de Compostela in northern Spain. The train was travelling at more than twice the permitted speed limit of 80 kmh (50 mph). Eighty people were killed, and 144 were injured. The horrifying derailment, subsequent carnage and twisted carriages were captured by security cameras on the route and widely broadcast in the media. The driver, with 30 years of experience under his belt, was reportedly on the phone talking to colleagues just before the crash.[2]

Loss of Situational Awareness

Although it was a work call, it no doubt contributed to the driver's loss of situational awareness. This is a circumstance similar to the one we may find ourselves in if using a mobile phone in a car, a subject we will explore in greater depth later on in the chapter. On the 424 km (263 m) long high-speed line in question, there were 31 tunnels and 38 viaducts.[3] Where a train is constantly running in and out of tunnels, a driver may also start to find it difficult to know exactly where he or she is on the route. Lineside signals and speed signage can become a mere blur, rendering them very difficult to interpret. Of course, the whole situation may be exacerbated many times over if a driver has diverted a large part of their attention to a phone call.

Just imagine travelling down a monotonous stretch of motorway in your car at twice the speed limit and attempting to read the signage. This will give you some idea of the additional difficulties faced in assimilating safety information at high speed, whilst retaining a sense of where you are physically. Your brain will inevitably start to struggle the faster you go. Now add a phone call into the mix! In fact, this was why a special screen to text alerts to the drivers was installed in the train driver's cab. If these alerts had been functioning, the driver would have received a text message several kilometres before the approaching curve. A short while later, there would have been a message with a difficult-to-ignore yellow flashing frame, along with an audible signal.

The driver would then have been asked to acknowledge the message by tapping on the screen. If there were no response, the brakes would have been applied automatically after five seconds. Crucially, this system, which had been designed to prevent a major accident, had been switched off. It had been developed especially to restore a driver's situational awareness, thus ensuring a clear focus on the thing that mattered the most – keeping the train on the rails and all the passengers safe. In its absence, succumbing to distraction was always likely to cause an accident.

Organisational Factors in Brief

The driver's actions ultimately may have caused the accident, but organisational factors played an important role too. A full M4 analysis of the unfolding events in Spain would pay greater attention to these factors too. Here they are in brief:

- The onboard European Train Control System (ETCS) designed to assist the driver had been switched off the year before the crash. Consequently, there were no driver alerts as the train approached the sharp curve. If the ETCS had been switched on, the driver would have been alerted 4 km (2.5 m) before the curve.[4]
- There was no automatic braking on this particular section of track, although this had been installed on most of the high-speed tracks in the region.
- There was a fundamental lack of risk management. An accident of this nature was likely to occur around every 6 months, as estimated by the judicial investigation report.

MINDFULNESS FOR TRAIN DRIVERS

Driving a train can be very solitary work, with limited opportunity for social interaction. For drivers, this can prove to be both monotonous and stressful, providing the perfect conditions for mindlessness to take hold. As we have seen, it can contribute to distraction and eventual loss of situational awareness. In the wake of the derailment at Santiago de Compostela, there was a clear desire to prevent such a massive loss of life happening again in Spain. There is a common tendency these days to search for technological solutions to eliminate the risk of such an accident, but interventions focused on direct training for enhanced concentration are much rarer.

In one innovative study carried out in Catalonia, drivers working for the publicly owned operator FGC in Spain were trained in mindfulness.[5] FGC operate over urban, suburban and regional lines, carrying 80 million passengers a year. They identified that preventing loss of attention and staying alert were fundamental to the task of driving. Preferring to invest on this occasion in a technology-free intervention, they were brave enough to try something different. Twenty-three drivers from FGC took part in a Mindfulness-Based Stress Reduction programme, which was custom designed to improve concentration and attention. Drivers were trained to develop more resilience, cope better with stress, be more present in the cab and avoid being on autopilot.

After the course, 85 per cent of the drivers said their attention was 'higher' or 'much higher'. Other benefits they cited included improved concentration, better emotional regulation and resilience and the development of individual coping strategies for stress and fatigue. The results of this study were remarkable in other ways too.

Immediately prior to the course, the drivers had been involved in serious safety incidents, such as Signals Passed at Danger (SPADs) and station overruns. In the six months after the course, many of these drivers were completely incident-free. Ideally, one would want to see more data from a much larger sample and a follow-up study. Nevertheless, it highlights the potential power of mindfulness in reducing serious incidents. Some more comprehensive examples are provided in Chapter 12 on training interventions.

ROAD SAFETY AND DISTRACTIONS

There is far more research on the impact of distractions on car drivers than there is for train drivers. For example, a UK study observed 11,000 drivers on the roads, finding that as many as one in six of them were engaged in a distracting activity, such as smoking or talking on the phone or to a passenger.[6] Younger drivers were more likely to be distracted checking their messages, social media accounts or paying attention to their passengers.

The consequences of being distracted at the wheel under quite ordinary driving conditions can be unimaginably horrific. In one particularly tragic crash, a FedEx lorry driver on a hands-free call ploughed into the back of a mini-bus on the M1 motorway. Eight people were killed and four were left seriously injured.[7] One

TABLE 6.1
Contribution of Inattention to Accidents (US Data)

Source	Type of Data	Per cent
National Highway Safety Administration[8]	Police-reported crashes	25
National Accident Data[9]	Highway accidents	35–50
Field Experiment Date[10]	100-car driving study of crashes	78

four-year-old girl was left orphaned. The driver had been talking about football and Donald Trump for up to an hour with a fellow driver. Even with up to 11 seconds to respond to the stopped mini-bus with its hazard lights on, he had been unable to apply the brakes in time. This heartbreaking crash highlights the dangers of being on autopilot.

Just how big an issue is inattention in crashes in general? Unfortunately, the estimates can vary wildly, as depicted in Table 6.1.

With the lowest estimate still standing at 25 per cent, we can safely conclude that inattention is a major factor in crashes. The range of estimates is likely to be attributable to the way in which inattention is classified. Higher estimates may be due to using inattention as a 'catch all' category to cover a range of phenomena, such as fatigue, in-vehicle distractions and a variety of eye glances away from the road.

Watch Out for Distractions!

There are many distractions that can affect our attention, both inside and outside the car:

- Physical
- Visual
- Auditory
- Mental.

Physical. Apart from spreading crumbs and liquid over your car seats, eating and drinking is going to divert your attention from the road ahead. Smoking won't help much either. No one wants to be frantically looking for a burning fag end accidentally dropped in a moving car. And typing a text message is perhaps the most dangerous thing you can do when driving – avoid it at all costs.

Visual. Reading your smartphone's screen is a major distraction, but picking an inopportune moment to read the car's satellite navigation system or infotainment display may be just as bad. Outside the car, have you ever noticed how people slow down to see the aftermath of a collision? Be careful not to become the next accident statistic yourself. And scantily clad pedestrians may provide summer eye candy for some, but would you really want to explain that to the police after you skid off the road?

Auditory. Distractions of an auditory nature might be unseen, but they are potentially just as lethal as the things that are visible. Listening intently to someone on the phone, even when it is hands-free, can be a drain on your attention. Loud music can affect your driving style too. Heavy metal will likely encourage you to put the pedal to the metal, creating an unhealthy dislike of traffic lights in the process.

Mental. This fact is often forgotten, but if our mind starts to wander, we need to bring it back to the road. We might be thinking about something else, someone else, or an argument or conversation from earlier on. Those silent ruminations can be deeply distracting, using up mental resources that would be better employed for the task at hand. Driving is an enjoyable experience in its own right, there's no need to overlay it with personal baggage.

If we can limit the distractions laying claim to our attention, we are also freer to attend to road users outside the car, who may themselves be distracted. The unpredictable actions of vulnerable road users (such as small children and the elderly), cyclists and jaywalkers with headphones on, may all require our quick reactions.

One fruitful line of research has sought to compare the relative impact of different types of distraction on drivers. Not all distractions are equal in their ability to suppress brain activity for safe driving. Some studies have taken direct measures of electroencephalographic (EEG) brain activity, providing some compelling evidence. Drivers who talk on their mobile phones experience a loss in their information processing capacity.[11] In other words, their attention is being diverted elsewhere. Listening to the radio or an audiobook is not nearly as distracting. The greatest danger actually comes from speech-to-text interactions with email, which can increase mental workload to unsafe levels whilst driving.

PERCEPTUAL BLINDNESS

Though individuals will respond to distractions in unique ways, the consensus is that they impede a driver's ability to spot hazards and react in time. Distractions can overload our attentional resources, depending on their intensity, duration, frequency and our own susceptibility to them. Even experienced drivers can suffer from a kind of 'perceptual blindness'. You often hear the phrase that a driver 'looked but did not see'. On the roads, for example, we pay measurably less attention to motorcyclists than we do to 'fatter' cars. The brain processes larger objects faster because they take up more visual space in our field of vision.[12] A short glance down the road may not be enough to clock a 'skinnier' motorcycle speeding towards you. So why do not we visually scan the road for longer?

Expectations tend to prime our attention as much as anything else. Motorcyclists only represent one per cent of traffic, though they make up 21 per cent of UK road deaths. The higher risk is also reflected in statistics from across Europe, America, Australia, China and New Zealand. Seeing a motorcyclist is a relatively novel experience in the vast majority of places. To mitigate the safety risk, we will have to look out for a hazard that isn't present for much of the time. In other words, expect the unexpected.

Beyond the convenient soundbite, however, there is a practical solution. Remember how M4 puts direct, first-hand experience at the heart of the approach? Direct experience is a fundamental ingredient for increasing risk awareness, shifting perspectives and effecting personal change.

Shifting Perspectives to Save Lives

In some fascinating research, those who drove cars and rode motorcycles were found to have higher levels of empathy and lower negative attitudes towards riders than all other car drivers.[13] These 'dual drivers' could mentally switch perspectives from driver to rider – and back again. Those participants who did not have the experience of both transport forms would naturally find this more difficult, though highly experienced drivers also showed greater empathy.

The novice drivers who participated in the study erroneously thought that motorcyclists could easily swerve to avoid obstacles. Their slightly elastic view of the laws of physics would be sorely tested on the roads in a real-life encounter with a biker. This all suggests that if you want to reduce the disproportionately high number of motorcyclist deaths on the roads, you need to shift perspectives for greater empathy and fewer accidents. We could all jump on a motorcycle to gain the necessary experience on the roads. Failing that, even going for a ride on a regular bicycle could help achieve the same effect and fundamentally change attitudes and pay road safety dividends.

Mobile Phones

Mobile phones can be lethally distracting whilst driving. For this reason, over 40 countries have passed laws to restrict the use of handheld mobile phones on the road. But catching distracted drivers isn't as easy as it may appear, because drivers will often hide their phones on their laps. In Canada, the police have taken a novel approach and aim to catch offending drivers by riding public transport.[14] Being higher up and able to look out of large windows on buses allows them to see right into cars. They have sound reasons for wanting to tackle the distraction epidemic.

Drivers are four times as likely to be involved in an accident when using a mobile phone, and this applies whether we are talking about handheld mobiles or hands-free.[15] Mobile phone use also has a measurable impact on our reaction times. One systematic review of 33 studies found that reaction times to events, such as the car in front's brake lights, increased by 250 ms compared to no-phone control conditions.[16] That may not sound like much, but for a motorist doing 50 mph (80 kmh), it translates into an additional 18 ft (5.5 m) before stopping. That could easily be the difference between stopping short of a child and a fatality. On one simulated driving task, participants on mobile phones could accurately recall far fewer of the objects they had encountered in a memory test.[17] Not being able to reliably recall whether one has seen a pedestrian, or an advertising hoarding, is a little disturbing. This is another kind of perceptual blindness, but one that should in theory be easier to control than in the motorcyclist example cited above. After all, we can choose not to use our phones.

Drivers tend to assume that hands-free phones leave their concentration largely unaffected, but the evidence says otherwise.[18] It seems that hands-free users do not use their eyes nearly as well as non-phone users. They were less likely to see peripheral objects in one research study that tracked eye movements, which suggests a kind of 'tunnel vision' that prevented them from seeing potential hazards by the side of the road.

And having a conversation with a friend on a hands-free phone results in a 50 per cent decline in navigation accuracy – you'll miss your turning far more easily this way.[19] A total of 14 US states have laws prohibiting a driver's use of handheld mobile phones. In contrast, all states allow the use of hands-free devices, despite research highlighting the equivalent safety risks. This appears to be a rather serious anomaly – why not extend the laws to apply a blanket ban to all types of mobile phone use in vehicles? In the absence of targeted hands-free legislation, we will have to fight our potentially dangerous attachment to our phones, whether or not our fingers are wrapped around them.

Is Electronic Autopilot Any Safer?

We have seen how inattention associated with going on autopilot can have tragic consequences. Our brains suffer an information-processing deficit and we fail to notice hazards, such as approaching motorcyclists or small children by the roadside. We also miss our turnings and end up someplace else. In traffic, we may fail to respond altogether to brake or hazard lights in our field of vision. So why not just hand over control to a machine?

Tesla offers a technologically sophisticated autopilot on their cars, a feature designed to make journeys less stressful. It can effectively take control of the car for highway driving. The corporation is not wrong to point out that, statistically speaking, you are more likely to die behind the wheel of one of their cars due to driver error than computer error. On average in America, there is a fatality once every 94 million miles. However, it was only after 130 million miles that a Tesla driver was killed, ultimately by computer code that failed to read the road conditions accurately enough.[20] A 40-year-old Tesla enthusiast was at the wheel. His car attempted to drive at full speed under the trailer of an 18-wheel truck crossing the highway. The Tesla's sensors could not correctly distinguish between the bright sky and the large white truck. Though statistically unlikely, such accidents still have the power to throw the future of autonomous driving into disarray. They raise searching questions about electronic autopilot modes. The implications for fallible human drivers are unlikely to be resolved anytime soon.

In fact, the 'ironies of automation', as they have been called, can be traced back many decades.[21] The chief irony is that automatic control systems are installed because it is assumed they can do the job better than a human operator. Yet the operator is still asked to make sure everything is working effectively. For this reason, Tesla's autonomous software 'nudges' drivers to keep their hands on the wheel to ensure they are paying attention. Tesla admits that "Autopilot is getting better all the time, but it is not perfect and still requires the driver to remain alert." The question is: how long can a driver reasonably stay alert for?

'Vigilance' studies suggest that it is impossible for a human being, even a highly motivated one, to pay effective visual attention to a source of information for more than 30 minutes.[22] That is not very long, especially if you are driving down a long stretch of monotonous highway. After that time, it becomes humanly impossible to monitor the traffic for abnormal situations, such as an 18-wheel truck crossing your path in bright sunlight. The situation called for the Tesla driver to be quicker than the computer in detecting and responding to an imminent crash – yet another irony.

A misplaced faith in the power of autopilot technology could increase the likelihood of a crash in some situations, at least if we fail to stay alert. Examples of this abound on the internet. Tesla owners can be seen taking naps in busy traffic, or even playing air guitar to Billy Idol. Whether or not you like the lyrics to his song Rebel Yell, this is one activity you should clearly not be doing more, more, more! Keep your hands on the wheel instead.

Uber's Fatality

The unfolding saga of self-driving cars continued when an Uber car failed to detect a 49-year-old pedestrian crossing the road with a bicycle at night in Arizona.[23] The car was moving in autonomous mode. The police video footage shows the Uber car careering into the woman, even though she is clearly visible. Uber's Lidar technology, which uses lasers so the car can 'see' the world around it, was fiercely criticised by experts. Neither the darkness nor shadows should have prevented the car from seeing the woman directly in its path. The video footage also shows how the 'safety driver' inside the car failed to see the woman until it was too late. As in the Tesla example, this driver, who was meant to be monitoring the automated system, is likely to have become bored or disengaged. Though she had about two seconds to react, she was no longer alert enough to prevent the accident.

An alert driver may have been able to brake or swerve in that time. Once again, this highlights the pitfalls of humans overseeing automated systems. When human drivers are called back 'into the loop' to deal decisively with a hazard or obstacle, their brains may have lost the ability to respond effectively many minutes earlier. Drivers inevitably suffer lapses in concentration, especially if there is little to keep them engaged. It is a risky endeavour to put them in charge of highly automated cars, whilst expecting them to override flawed hazard detection systems.

INTERRUPTED BRAINS

> In a city of the future, it is difficult to concentrate.
>
> – **Lyrics to Palo Alto**, *Radiohead, 1998*

In the time it takes you to read this section, there is a high chance your brain will be interrupted by your phone buzzing. Even if that isn't the case, you may be seeking out distractions purely out of habit. As we noted in Chapter 4, the brain can be hijacked as it searches for its next dopamine hit. A former vice-president of user growth at Facebook went as far as saying:

> The short-term, dopamine-driven feedback loops that we have created are destroying how society works. No civil discourse, no cooperation, misinformation, mistruth.[24]

This feeds into a societal narrative about our distracted selves, but the effects can also be felt in the workplace, because they impact on productivity levels. Whenever we attempt to do two or more things at once, we end up multi-tasking and dividing our attention. These days, we often consume media simultaneously on our smartphones, and on a range of other devices, such as tablets, computers and TVs. Toggling between devices in this way is called 'media multi-tasking'. You may well have experienced watching your favourite programme on TV whilst your attention is constantly diverted to your smartphone buzzing with notifications. Before you know it, you've reached the end of an episode, but you can't recall much of it at all. In fact, media multi-tasking may well be robbing us of brainpower.

One carefully designed Stanford University study researched the effect of multi-tasking on our cognitive capacities.[25] It began by identifying 12 different media forms: print, television, computer-based video (such as YouTube and online TV), music, non-music audio, video games, telephone and mobile voice calls, instant messaging, SMS (text messaging), email, web-surfing and computer applications (such as word processing). To determine the level of media multi-tasking, a questionnaire was administered by the experimenters, who were then able to identify two distinct groups: chronically heavy and light media multi-taskers. And you'd be wrong to think that heavy media multi-taskers were better equipped for dealing with distractions in everyday life.

In some ways, our brains are like computers. Having too many computer applications running in the background slows everything down. In the same way, simultaneously concentrating on too many tasks slows down our brains, just as it does when you drive and engage in conversation. With computers, though, you can normally close those applications down again to restore processing power. It is a little different for 21st-century human brains. Unfortunately, processing multiple information streams in a near-permanent state of distraction can fundamentally alter our processing capabilities. The results from the study are rather disturbing, with clear implications for consumers who passively feed on lots of media.

Chronically heavy media multi-taskers have greater difficulty in focusing their attention. They also struggle more to filter out irrelevant stimuli from their environment. And they are less effective at suppressing the activation of irrelevant tasks. In the face of distractions, light media multi-taskers find it easier to focus on a single task. This was all borne out by the scientific evidence. Heavy media multi-taskers were 426 ms (almost half a second) slower than light media multi-taskers at switching to new activities. For a car travelling at a speed of 30 mph, this would equate to covering an additional 5.7 m (over 18 ft), which could be the difference between life and death if the driver responded late to a hazard in the road. For new sections of the same activity, heavy media multi-taskers were also 259 ms slower to engage in the experiment.

The Stanford University study raises profound questions about the nature of human cognition in the future. Millennials, in particular, are more vulnerable and susceptible to constant media bombardment. They have never experienced adult life in the pre-digital era, but they are also more likely to be facing workplace challenges with diminished cognitive capacities. Young minds will need a high degree of self-awareness to notice the influences they are being subjected to.

Mindfulness not only provides an antidote to a world full of distractions, but it can also train minds for enhanced attention.

TAILORED M4 TRAINING FOR IMPROVED HEALTH, WELLBEING AND SAFETY

The specific application of mindfulness with a tailored programme for safety-critical workers is unique, with the potential to surpass most conventional efforts to effect behavioural change. Conventional efforts often fail to achieve their desired objectives, largely because they attempt to modify people's behaviour with a rather straightjacketed idea of how their behaviour should look. A behavioural approach might be fine when we are training rats, pigeons or dogs in the Pavlovian tradition, but human beings are far more complex.

Making safe decisions in complex environments involves a high degree of self-awareness.

An array of cognitive skills must come together. Vigilance, judgement, reasoning, memory and learning are all exercised. Mindfulness training teaches people to think on their feet and adapt to novel situations, enhancing their ability to respond safely in all situations, while fully aware of the dangers they face in real time. It promotes risk awareness to change safety related behaviour on the ground, whilst contributing to everyone's wellbeing in a much more holistic sense.

Uniquely, it helps prevent people's minds from wandering, improving their ability to focus and concentrate. Its application to high-hazard industries will have far reaching effects for health, wellbeing and system safety.

If you would like to greatly reduce the likelihood of a safety incident, give your business an edge and implement a training programme designed with the needs of your people in mind, visit www.m4initiative.com for more details.

KEY POINTS

- Distraction and a loss of situational awareness can be a major contributor to major accidents, such as the derailment at Santiago de Compostela in Spain.
- Mindfulness can be used to counteract the effects of driving on autopilot. It enhances alertness, concentration and emotional resilience.
- By building self-awareness, mindfulness has the potential to produce safer workplace behaviours, where conventional behavioural approaches are often a struggle.
- Road users can also apply mindfulness techniques. Distraction contributes to at least 25 per cent of crashes, but it is true contribution may be much higher.
- By shifting perspectives, the associated gain in empathy can lead to safer driving.
- Automation has its ironies. The electronic autopilot systems installed in some vehicles carry their own safety risks, requiring a human being to intervene at critical moments.
- A constant, high level of media multi-tasking can diminish our cognitive capacities over the long term, affecting our ability to focus our attention.

M4: APPLYING MINDFUL SAFETY TO BEAT DISTRACTION

INDIVIDUAL

- Notice how different types of distractions affect your alertness, vigilance and concentration.
- If your mind starts to wander, bring it back to the present moment. Mindfulness has specific techniques for practising this.

RELATIONAL

- If you are a car driver, be alert to perceptual blindness and give yourself time to look out for more vulnerable road users.
- Be aware that hands-free calls are just as distracting as those taken on a mobile device. Consider banning yourself from all in-car calls.
- Practise taking on board other road users' perspectives. Consider riding a bike or a motorcycle to shift your own perspective and become a better driver.
- If you are a heavy user, cut down on media multi-tasking. In general, pay attention to how multiple media streams may affect your performance.
- Does your partner, friend or colleague say you seem constantly distracted? Is a high consumption of media from various sources affecting your relationships at home or in the workplace?
- Observe whether distractions are affecting the behaviours of others. You may notice fidgeting, absent mindedness, or an inability to stay on task. Start a conversation with them about the distractions in their lives. It is a stimulating topic for discussion – just do not get distracted!

ORGANISATIONAL

- How does your organisation manage workplace distractions? This is one area where there are usually gaps. Here are some questions worth asking:
- Is the workforce taught how to manage interruptions to their work?
- Are workers aware of the risks of having to monitor automated machinery for longer than 30 minutes?
- Can the IT department either switch off or limit the stream notifications on computers and mobile devices by default?
- Is mindfulness taught as a distraction-busting technique?

SOCIETAL

- Lobby for vehicle manufacturers to fully test distracting technologies, such as speech-to-text for email, before releasing it onto the market. Provide feedback to manufacturers if such technology proves to affect people's ability to concentrate.
- Lobby for changes in the law.

NOTES

1. Crawford, M. (2015). *The World Beyond Your Head. How to Flourish in Age of Distraction*, p. 4. London: Penguin Books.
2. Dawber, A. (2015). Spanish train crash: driver facing 80 homicide charges, but rail bosses cleared. *The Independent.* 8 October 2015. Available from http://www.independent.co.uk/news/world/europe/spanish-train-crash-driver-facing-80-homicide-charges-but-rail-bosses-cleared-a6686951.html (Accessed 02.03.2018).
3. Kemp, R. (2013). *The Spanish Train Crash – A Deficit in Situational Awareness?* University of Lancaster. 13 August 2013. Available from http://www.lancaster.ac.uk/news/blogs/roger-kemp/the-spanish-train-crash--a-deficit-in-situational-awareness/ (Accessed 02.03.2018).
4. Puente, F. (2015). ETCS: a crucial factor in Santiago accident inquiry. *International Rail Journal.* 7 April 2015. Available from http://www.railjournal.com/index.php/signalling/etcs-a-crucial-factor-in-santiago-accident-inquiry.html (Accessed 02.03.2018).
5. Juncadella, O. (2016). *Using A Human Factors Approach to Improve Reliability and Safety.* Presentation given to European Rail Safety Forum, 27 January 2016.
6. Sullman, M., Prat, F. & Tasci, D.K. (2014). A roadside study of observable driver distractions. *Traffic Injury Prevention*, Volume 16(6), pp. 552–557. doi:10.1080/15389588.2014.989319.
7. M1 crash deaths: driver in fatal crash on hands-free call. *BBC News.* 28 February 2018. Available from http://www.bbc.co.uk/news/uk-england-beds-bucks-herts-43205492 (Accessed 10.03.2018).
8. Ranney, T., Mazzae, E., Garrott, R. & Goodman, M. (2000). *NHTSA Driver Distraction Research: Past, Present and Future.* Available from https://www-nrd.nhtsa.dot.gov/departments/.../driver-distraction/PDF/233.PDF (Accessed 10.03.2018).
9. Sussman, E.D., Bishop, H., Madnick, B. & Walker, R. (1985). Driver inattention and highway safety. *Transportation Research Record*, Volume 1047, pp. 40–48. Available from http://onlinepubs.trb.org/Onlinepubs/trr/1985/1047/1047-007.pdf (Accessed 10.03.2018).
10. Dingus, T.A., Klauer, S.G., Neale, V.L., Petersen, A., Lee, S.E., Sudweeks, J. & Knipling, R.R. (2006). *The 100-Car Naturalistic Driving Study: Phase II. Results of the 100-Car Field Experiment* (DOT HS 810 593). Washington, DC: National Highway Traffic Safety Administration. Available from https://www.nhtsa.gov/DOT/NHTSA/NRD/.../Driver%20Distraction/100CarMain.pdf (Accessed 10.03.2018).
11. Strayer, D.L. & Drews, F.A. (2007). Cell-phone-induced driver distraction. *Current Directions in Psychological Science*, Volume 16, pp. 128–131.
12. Crundall, D. (2017). *What motorists know and don't know (about motorcyclists).* Motorcycle Safety Messages. MS-03. 27 March 2017. Available from: https://lifesaversconference.org/wp-content/uploads/2017/04/CrundallD2.pdf (Accessed 03.03. 2018).
13. Ibid.
14. Nanowski, N. (2016). The better way to target texting drivers: York region police ride the bus. *CBC News.* 22 December 2016. Available from http://www.cbc.ca/news/canada/toronto/distracted-driving-police-tickets-york-region-1.3907419 (Accessed 04.03.2018).
15. Strayer, D.L., Watson, J.M. & Drew, F.A. (2011). Cognitive distraction while multitasking in the automobile. *Psychology of Learning and Motivation*, Volume 54, pp. 29–58.
16. Caird, J.K., Willnes, C.R., Steel, P. & Scialfa, C. (2008). A meta-analysis of the effects of cell phones on driver performance. *Accident Analysis and Prevention*, Volume 40, pp. 1282–1293.
17. Strayer, D.L. & Drews, F.A. (2007). Cell-phone induced driver distraction. *Current Directions in Psychological Science*, Volume 16, pp. 128–131.

18. Strayer, D.L., Watson, J.M. & Drew, F.A. (2011). Cognitive distraction while multitasking in the automobile. *Psychology of Learning and Motivation*, Volume 54, pp. 29–58.
19. Strayer, D., Turill, J., Cooper, J.M., Coleman, J.R., Medeiros, N. & Biondi, F. Assessing cognitive distraction in the automobile. *Human Factors*, Volume 57(8), pp. 1300–1324. Available from http://journals.sagepub.com/doi/pdf/10.1177/0018720815575149 (Accessed 10.03.2018).
20. Yadron, D. & Tynan, D. (2016). Tesla driver dies in first fatal crash while using autopilot mode. *The Guardian*. 1 July 2016. Available from https://www.theguardian.com/technology/2016/jun/30/tesla-autopilot-death-self-driving-car-elon-musk (Accessed 10.03.2018).
21. Bainbridge, L. (1983). Ironies of automation. *Automatica*, Volume 19(6), pp. 775–779. Available from: https://www.ise.ncsu.edu/wp-content/uploads/2017/02/Bainbridge_1983_Automatica.pdf (Accessed 11.03.2018).
22. Mackworth, N.H. (1950). Research on the measurement of human performance. Reprinted in Sinaiko, H.W. (Ed.), *Selected Papers on Human Factors in the Design and Use of Control Systems* (1961), pp. 174–331. New York: Dover Publications.
23. Levin, S. (2018). Uber crash shows 'catastrophic failure' of self-driving technology, experts say. *The Guardian*. 22 March 2018. Available from https://www.theguardian.com/technology/2018/mar/22/self-driving-car-uber-death-woman-failure-fatal-crash-arizona (Accessed 05.08.2018).
24. Wong, J.C. (2017). Former Facebook executive: social media is ripping society apart. *The Guardian*. 12 December 2017. Available from https://www.theguardian.com/technology/2017/dec/11/facebook-former-executive-ripping-society-apart (Accessed 18.03.2018).
25. Ophir, E., Nass, C. & Wagner, A.D. (2009). Cognitive control in media multitaskers. *PNAS*, Volume 106(7). Available from http://www.pnas.org/content/pnas/106/37/15583.full.pdf (Accessed 18.03.2018).

7 The Mental Health Elephant at Work

There is no health without mental health; mental health is too important to be left to the professionals alone, and mental health is everyone's business.

– Vikram Patel

This chapter is all about mental health in the workplace and what mindfulness can do to help. It therefore attends most to the individual and organisational levels within the M4 approach. I often meet people in the consulting room who feel helpless and at the end of their tether. Here, I write from my knowledge base as a practising, registered counsellor and mindfulness teacher. However, I do not go into a huge amount of detail about clinical conditions. If you are suffering from poor mental health, it is important to seek the help of a professional.

Mindfulness has a role to play in sustaining good mental health, but it is not a panacea for all mental health problems. Psychological and emotional distress can manifest in symptoms of stress, anxiety and depression. Focusing on the proactive side of dealing with these symptoms places a natural emphasis on developing self-awareness, inner calmness and practising daily meditation. And the research shows it can be highly effective.

The tragic consequences if mental health problems go unchecked for a long time are all too clear to see. I will start here by sharing the case of the Germanwings pilot who committed suicide in the cockpit. Crashing his plane into the Alps, he took all the crew and passengers with him. This is a tough subject matter, but if the lessons can be adhered to we can look to the future with far more optimism. There is no health without mental health, but there is no safety without it either.

Case Study: Mental Health Lessons from Germanwings 9525

When 34-year-old Captain Patrick Sondenheimer took the controls of Germanwings flight 9525 on 24 March 2015, nothing could have prepared him for the terrifying situation he would face on the way from Barcelona to Dusseldorf. Sondenheimer had ten years' flying experience under his belt. To spend more time with family, he had recently switched from long-haul to short-haul flights. His co-pilot was 27-year-old Andreas Lubitz. Lubitz had a history of mental illness and was suffering psychotic symptoms.

On the Day of the Accident

The Airbus 320 takes off from Barcelona at 9:01 am and begins travelling over the sea towards France. It takes about half an hour to climb to the cruising altitude of

38,000 ft (11,600 m). Lubitz is initially courteous towards Captain Sondenheimer. He becomes a little curt when the captain gives the mid-flight briefing on the planned landing.[1] With the benefit of hindsight, that is perhaps the only inkling of how events would unfold.

At 9:30, the plane has its final contact with air traffic control. It is just a routine communication about permission to continue on route. The captain informs Lubitz he is leaving the cockpit, most probably for a toilet break, and asks him to take over radio communications. The cockpit door is heard opening and closing on the voice recorder. Seconds later, Lubitz seizes the opportunity to manually put the aircraft into a descent from 38,000 to 100 ft. The plane begins plummeting through the sky at nearly 4,000 ft a minute.

From 9:33 on, air traffic controllers try to contact the pilots. There is no response. Lubitz does not say a word, though his breathing remains normal. Noises similar to a person knocking on the cockpit door can be heard. There are some muffled voices and then an audible request for the cockpit door to be opened. Despite repeated attempts by crew members and air traffic control to get Lubitz to respond, nothing works. Noises similar to violent blows on the cockpit door are recorded on five instances in the last 90 seconds of flight. The 'Terrain, Terrain, Pull Up, Pull Up' warning is sounded around 30 seconds before the final impact at 9:41. Passengers can be heard screaming only near the very end.

At the hands of one suicidal pilot, the plane crashes into the French Alps at 430 mph (692 kmh). Death is instant for all 144 passengers and the six crew members.

Findings from the Safety Investigation

The BEA (the French Civil Aviation Safety Investigation Authority) produced a final report of its findings in March 2016. The BEA's primary aim is to prevent accidents and incidents.[2] The 110-page report makes a grim read, but it is by no means easy to see how a tragedy like this could be prevented in the future, even after their thorough investigation and insight into the causes.

The immediate cause of the crash was Lubitz's decision to commit suicide while alone in the cockpit. He had effectively hidden a psychiatric condition that made him unfit to fly. His mental state in the months leading up to that fatal day went unnoticed by the pilots that flew with him. He had not sought any support through pilot support programmes available to Germanwings pilots, such as the well-established Mayday foundation designed to help those experiencing personal difficulties. The BEA concluded that:

> No action could have been taken by authorities and/or his employer to prevent him flying that day, because they were informed by neither the co-pilot himself, nor anybody else, such as a physician, a colleague, or family member.[3]

This goes right to the heart of the matter. How can an accident like this be prevented where the information required to intervene is not available to the authorities in the first place? The authorities face similar scenarios in trying to prevent a terrorist attack on the public by a lone gunman. Without the availability of reliable intelligence, it is extremely difficult to take pre-emptive action. It is especially difficult if the perpetrator goes to great lengths to hide the threat he poses to the public.

Lubitz started to show symptoms possibly associated with a psychotic depressive episode in December 2014. This was five months after the last revalidation of his class 1 medical certificate, which all airline pilots require in order to exercise their license. He consulted various doctors and was prescribed anti-depressant medication by the psychiatrist treating him. EU regulations stipulate that pilots should seek the advice of an Aero Medical Examiner (AME) after starting the regular use of medication. Lubitz, however, neglected to tell any AMEs what he was taking and continued to fly until the day of the accident.

In February 2015, Lubitz consulted a private physician. A possible psychotic depressive episode was diagnosed and he was referred to a psychotherapist and psychiatrist. In March 2015, just two weeks before the accident, the same physician was concerned enough to recommend psychiatric hospital treatment.

A History of Depression

Lubitz's mental health problems can be traced further back. He dropped out of his training in November 2008 to be treated by a psychiatrist who prescribed medication. In July 2009, the depressive episode was declared over. Lubitz had recovered. The Aero-Medical Centre issued his class 1 medical certificate, but with a waiver stating it would be invalidated if there were any relapses into depression. All the AMEs who examined him between 2010 and 2014 were aware of his medical history. Their professional evaluations of his psychological fitness deemed him fit to fly. His class 1 medical certificate was therefore revalidated on each occasion without the need for further examination.

One could argue that given Lubitz's history of depression, more extensive psychiatric evaluations should have occurred in the interests of protecting the public. However, the BEA contacted specialists in aerospace medicine and psychiatrists who generally agreed that:

> Detection tools and methods can remain ineffective in cases where the patient is intentionally hiding any history of mental disorder and/or is faking being in good health. This is why most believe that putting in place extensive psychiatric evaluation as part of routine aeromedical assessments of all pilots would not be productive or cost effective.[4]

It is acknowledged, though, that in the case of individuals with a history of mental illness, the process could be strengthened with more frequent and more thorough evaluations.

Grounds for Breaking Confidentiality?

None of the health professionals involved in Lubitz's treatment reported any public safety concerns to the authorities. Believing in the universally accepted principle of medical confidentiality, they upheld the trust between doctor and patient, as would be expected in the vast majority of cases. Those treating him would probably have been aware that he was a pilot. They could have reported their concerns, at least in theory, so what stopped them?

German regulations contain provisions to punish doctors breaching medical confidentiality, including possible imprisonment of up to a year. There was no formal definition of 'imminent threat' or 'threat to public safety' to guide them, and the fear of being sued for passing on private medical information may have weighed heavily on their minds. The balance between medical confidentiality and public safety is clearly a delicate one. In hindsight, it appears the balance was skewed in favour of medical confidentiality to the detriment of public safety. However, simply changing the regulations to allow for breaches of medical confidentiality under certain circumstances might have unfortunate side effects.

If you were aware your doctor could breach your medical confidentiality, would you be so candid about your mental health? A slightly paranoid, Kafkaesque-element might creep in. You might start fearing your information could be shared without your permission. This could drive the expression of mental health issues further underground, inadvertently increasing the stigma attached. Allowing doctors to pass on medical information anonymously may provide part of the solution, but this is not without its pitfalls either.

LEARNING FROM GERMANWINGS

The ultimate cause of the Germanwings accident was a distressed individual no longer in touch with reality. One cannot underestimate the stigma attached to his mental illness. The potential consequences of owning up to his problems probably felt insurmountable to Lubitz. Psychologically speaking, his career, indeed his whole existence, was under threat.

It would be complacent to assume such an accident couldn't happen again. It will always be prudent for organisations to assess the safety risk posed by an individual in such a state of mind. Early interventions should be considered wherever possible. The questions that will need answering are:

- Is the organisation confident that they will be able to detect a sick or unfit employee who poses a risk to public safety? If detected, how will this risk be effectively managed?
- Do current EAPs (Employee Assistance Programmes) provide robust enough support for staff with long-term mental health difficulties?
- How can we reduce the stigma attached to mental illness in the workplace for employees?

Fully addressing the stigma attached to mental illness will no doubt require some organisational soul searching. If we are truly interested in the health and wellbeing of employees who have a role in protecting the public from harm, we will need to be much more proactive.

What Was Going on in Lubitz's Mind?

We'll never know for sure what Lubitz was really thinking because he is no longer around to ask. Nevertheless, if we are interested in preventing a similar accident from happening again, we need to understand what may have been going on in his mind. Lubitz was probably suffering a psychotic depressive episode. He was obviously unfit to fly around the time of the accident. Such an episode is often accompanied by a loss of connectedness with reality, in addition to impaired judgement and decision-making.

What is clear is that Lubitz was stuck in a world that did not appear to offer any solution, or way out.

The late American suicidologist Edwin Shneidman suggested that a common feature of a suicidal state of mind is 'cognitive constriction'.[5] In essence, this means that a rigid, narrow pattern of thinking comparable to tunnel vision is present. A person in this state of mind finds it very hard to engage in effective problem-solving behaviour, seeing their options in 'all or nothing' terms. Such a person may be more vulnerable to suicidal thinking, especially if they have high standards and expectations. If failure or disappointment is attributed to their personal shortcomings, Shneidman suggested they may come to view themselves as incompetent, worthless and unlovable.

Airline pilots are very often passionate about flying and enjoy the high social status attached to their role. Lubitz's livelihood and professional identity were both under threat if he couldn't fly. There were also severe financial consequences, of around 60,000 Euros, if he lost his license. In these circumstances, he was also facing the loss of his future income. Losing his license would have brought his career to an abrupt end and effectively destroyed his professional ambitions. He may have felt that a graceful, professional exit from the airline industry was simply not possible.

WORKPLACE STRESS AND MENTAL HEALTH

There are clearly lessons to be learned from the Germanwings accident, and not just for people with severe mental health problems in safety critical environments. If we feel unable to disclose the true state of our mind to our employers, it is potentially a problem for us all. The fact is that we all struggle with our mental health sometimes. The oft-quoted figure is that one in four will suffer a mental health problem at some point in their lifetime. Quoted by mental health charities and government officials, the figure's origin is difficult to trace, and the evidence base is unclear.[6] Nevertheless, it helps concentrate minds on the prevalence of these issues.

One of the main drivers of mental ill health is workplace stress. This can lead to decreased employee engagement, suboptimal performance and high turnover rates. It may be costing the US economy alone $450 to $550 billion every single year.[7] According to one US survey of 17,000 employees, 63 per cent of respondents said that their workplace stress had a significant impact on their mental and behavioural health.[8,9] The results of what workers said when asked what symptoms of stress they had experienced in the last month are shown in Table 7.1.

Another survey of more than 2,000 American workers looked at a broad range of working conditions.[10] It provides some very useful context for the stress employees may be experiencing. Table 7.2 shows how pressurised and potentially mentally taxing people are finding the workplace.

TABLE 7.1
Symptoms of Stress in the Last Month for US Workers

Symptoms of Stress	Per cent
Irritability or anger	37
Nervousness or anxiety	35
Lack of interest or motivation	34
Fatigue	32
Feeling overwhelmed	32
Feeling depressed or sad	32

TABLE 7.2
Adapted from the American Working Conditions Survey (AWCS)

Type of Work Intensity	Per cent
Working at high speed (at least half the time)	66
Working to tight deadlines (at least half the time)	66
Working in one's free time to meet deadlines	50
Finding there is not enough time to do the job	27

Two thirds found that they were frequently working at a high speed (i.e. high work rate) or to tight deadlines. Half found that work spilled over into their free time when they had deadlines to meet. Just over one in four found they had too little time to do their job.

The Wall of Silence

Despite the obvious prevalence of workplace stress, there appears to be a wall of silence when it comes to talking about mental health issues at work. Employees are probably right to be wary – for those returning to work after suffering from mental ill health, there is a real risk of being stigmatised, or even suffering discrimination. Following an absence related to mental ill health, 32 per cent of UK employees felt their line managers treated them differently after returning to work. One in five said their colleagues' attitudes had changed towards them too. Some said they were walking on eggshells, without knowing how to approach them or what to say.[11]

No wonder employees find it difficult to openly disclose their mental health issues. If we perceive – rightly or wrongly – that our colleagues, or employer, will take a dim view of any issues, we are likely to put up our own walls of silence. People are often sceptical whether they'll be listened to compassionately at work, and may also fear being labelled.

No one wants to be viewed as having 'psychological problems' in the workplace. This creates a catch-22 situation. Many are afraid to discuss the subject at work, but this contributes to the masking of the problem. It is then difficult for employers, who remain unaware of the problem's scale, to tackle it effectively. Traditionally, EAPs, which often include counselling, aim to fill the gap. But simply 'outsourcing' mental health issues to a specialist provider may just be a form of corporate avoidance.

The M4 approach emphasises more attuned, mindful listening to our colleagues' struggles. Beyond a person's state of mind, and whether they are feeling 'anxious' or 'depressed', there will be a personal story, which we will probably be able to relate to. We needn't label it, but we can attempt to listen to it non-judgementally. Managers can encourage workers to talk more openly about what is really bothering them. In order to break down the wall of silence, the mental health elephant in the workplace needs to be fully acknowledged. The potential advantages are too good to ignore.

De-stigmatising mental health holds the promise of reduced rates of absence and staff turnover, and greater productivity.

MINDFULNESS TO OVERCOME MENTAL ILL HEALTH

In Chapter 4, we took a brief look at the health benefits of practising mindfulness. Here, we'll take a closer look at the research highlighting its role in fighting mental ill health and building resilience. On the day of the Alps crash, Lubitz clearly should not have been flying. He couldn't think clearly or make intelligent decisions. A toxic mixture of depression, stress and fear are likely to have shut down activity in his prefrontal cortex, the area of the brain responsible for executive functions such as decision-making. There can be no doubt he needed long-term clinical support to manage his condition – could a mindfulness intervention have made the difference?

In more everyday situations, long before the mental breaking point Lubitz reached, evidence suggests mindfulness can play a critical role in preventing mental ill health. One interesting study followed 22 people who participated in an eight-week Mindfulness-Based Stress Reduction Programme.[12] Both anxiety and depression were found to drop significantly amongst almost everyone participating; the frequency and severity of panic attacks also dropped. These benefits were sustained. Three months later the participants were still virtually free of panic attacks, and three years after that, most were still practising mindfulness in ways meaningful to them.[13] Though admittedly this study was based on a small number of individuals, and despite the fact that there was no control group for comparison either, it still demonstrates how mindfulness training can alleviate suffering.

For those seeking a higher standard of scientific evidence, much stronger empirical results were obtained from a ground-breaking study of 145 participants.[14] The majority of these (77 per cent) had experienced three or more depressive episodes. The Mindfulness-Based Cognitive Therapy (MBCT) programme administered in this case used a randomised control group for comparison. The results were staggeringly good. Those with a prior history of three or more depressive episodes – the especially 'stubborn' cases in other words – relapsed at half the rate of the control group. Half the rate! As the authors of the study explain:

> The focus of MBCT is to teach individuals to become more aware of thoughts and feelings and to relate to them in a wider, de-centered perspective as 'mental events' rather than as aspects of the self or as necessarily accurate reflections of reality.[15]

In this approach, the grip of depression is loosened in a way that may appear counter-intuitive at first. MBCT recognises that attempting to talk yourself out of depression may be futile, since seeing it as a problem to be solved can actually make matters worse. It is more advantageous to practise being aware and accepting thoughts and feelings as they are. Thoughts and feelings are treated as though they are clouds in the sky – they come and go and there is no special need to change them with conscious effort.

This way of thinking challenges the idea that the mind can be completely trapped by a consistently negative, fixed state. Not even clinically depressed people are depressed 100 per cent of the time. They have lighter, more joyful moments too. Encouraging them to pay close attention to the changing nature of their inner experiences can help their recovery enormously. It is not difficult to see why the UK's National Health Service recommends MBCT for people with a history of three or more episodes of chronic depression. However, the applications of this and mindfulness practice in general extend far beyond clinical settings.

Mindfulness is a proven tool for recovery, but it can also be used to build up resilience, as regular practice leads to greater activation of the pre-frontal cortex. This is associated with a reduction of fear, anxiety and aggression, and consequently there is less activity in the amygdala. This is associated with a reduction of fear, anxiety and aggression, and consequently there is less activity in the amygdala, the almond-shaped part of the brain that determines how we react to things. According to eminent researcher Richard Davidson, the left side of the pre-frontal cortex can be 30 times as active in a resilient person, compared to someone who is not.[16]

STRESS, MINDFULNESS AND CHOICE

How we handle stress determines whether events control our lives, or whether we can let things go to enjoy ourselves more fully. We can free ourselves from automatic, habitual ways of reacting by becoming more aware of our thoughts, feelings and bodily sensations. By learning to respond more skilfully to events, we can reduce the symptoms of stress, anxiety and depression.

We often find ourselves falling into one of the following ways of reacting to things:

- **Avoidance**. There is natural tendency to get rid of an experience we do not like.
- **Indifference**. We may switch from the present moment to go somewhere else in our minds.
- **Attachment**. We may attach ourselves to our present experience in an attempt to prolong it. When our present experience changes, we may have difficulty accepting it.

Each of these ways of reacting can cause problems. Avoidance is a particularly unhealthy strategy. The first essential step towards responding mindfully is to acknowledge how these reactions can affect us.

The Stress Reaction Cycle

Our automatic reactions can make stress worse, and simple problems can grow into larger ones. Over the course of our lives, reacting unconsciously to stress significantly increases our risk of both physical and psychological illness.

External Stressors and Internal Stress Events

We all experience external stressors that have their origin in our environment. These may be physical, social and economic. For example, mental or physical illness, an obnoxious boss, or financial strain all qualify in this respect. These stressors generate stress, changing our bodies and influencing our lives. But exactly how much stress they generate varies hugely from person to person. Inside our heads and bodies, our thoughts and emotions are strongly influenced by how we perceive these external stressors. Our own reactions to stress can have a considerable impact, affecting our immune, nervous, musculoskeletal and cardiovascular systems. The relationship between stressors, and the stress we experience in our lives, is unique to each person.

To illustrate the point, envisage yourself giving a presentation to your colleagues at work. For some, this can be a highly stressful event, evoking lots of anxiety. Others may eagerly await such an occasion because it provides an opportunity to demonstrate their knowledge, and they will feel a kind of alert calmness.

Chronic Stress and Acute Stress

What if we must deal with a stressor over a long period of time and cannot find a way of mitigating it? This is called chronic stress. An example would be taking care of a disabled family member whose condition will not change. In contrast, acute stress refers to stress experienced over a relatively short period of time. Working to a tight deadline would be a good example here. Whether stress is chronic or acute, each person will react in ways unique to them. This will largely depend on how they perceive them as threats, either to their wellbeing or sense of self. A whole range of reactions is possible. If no threat is perceived, the reaction may be minimal, but in a threatening, emotionally charged situation, an alarm reaction may be triggered.

Fight or Flight

If an alarm reaction is triggered, our body prepares itself for action. This is an unconscious, physiological fight-or-flight reaction designed for our protection in life-threatening situations. Just as it is in animals, the reaction is there to help us maintain or regain control. This leads to a state of physical and psychological hyperarousal. Our muscles tense, we experience strong emotions, there is a rapid cascade of nervous-system firings, and stress hormones such as adrenaline are released.

At the same time, we become very alert and attentive. Our heart beats faster, our blood pressure rises and blood is redirected from digestion to the large muscles in our legs and arms. We have that unnerving sensation of butterflies in the stomach. Chased by a bear, there is absolutely no point in digesting food anymore! Our very survival is at stake, and all our energy is therefore diverted for a fight-or-flight response. The autonomic nervous system is responsible for this activity. In life-threatening situations, the fight-or-flight reaction makes perfect sense. But it can become a burden if it goes into overdrive in situations where our life is not actually under threat.

Psychological stress comes from real or imagined threats to our social selves more than to our physical selves. We do not want hyperarousal to become a way of life. But patterns can be set if the mind and body's automatic reactions are left unchecked over time. Unfortunately, chronic muscle tension, a faster heart rate, a desire to flee or to get into arguments or fights can become the norm.

Internalised Stress Reactions

The fight-or-flight reaction can build up inside us. It may start to feel as though it is destabilising our place in the world, as well as our sense of self. And it may be socially unacceptable for us to express the associated feelings and emotions that come with it. A common reaction is to deny we are experiencing them at all. We suppress them, hiding our true state of arousal from others. Instead, we carry everything around with us, unable to let go.

In nature, facing or running from a foe would eventually allow us a form of release. We might successfully defend ourselves, or escape to safety, paving the way for a natural recovery. Where that release is not possible, we are left in a state of agitation. Our stress reaction can become internalised. High blood pressure, digestive problems, headaches, backaches and sleep disorders, as well as stress and anxiety, may all follow.

Finding Another Way

People often cope with stress in self-destructive ways. Maladaptive coping refers to coping strategies that provide temporary relief, but are in fact damaging in the long term. Such strategies help us tolerate stress, whilst providing the illusion of control. For example, workaholism and overeating may be effective in suppressing our true feelings, but they are unhealthy over time. Alcohol, caffeine, sugar, nicotine and prescription drugs may all provide short-term props, but they ultimately increase stress and encourage dependency or addiction. And left unchecked, exhaustion, burnout, physical illness and depression are the likely outcomes.

We want to avoid becoming caught up in a vicious circle, believing that is just the way life is. With the right tools, we can break the cycle of stress reactivity. Once we are aware of our own patterns, we are in a good position to challenge them. Mindfulness can give us new choices through harnessing the power of awareness to help us cope better with difficult personal circumstances. We can become aware of when to take time out for hobbies and what we must do to look after ourselves. The practice of mindfulness encourages the gaining of a higher perspective, helping us to cope with stress in more productive ways.

Five Stress Busters

These five tips for reducing stress can make a big difference in daily life.

1. **Take a deep breath**. Never underestimate the power of deep breathing, especially in stressful situations. If you can breathe fully, deeply and slowly, you will be increasing the supply of oxygen to every cell in your body. Deep breathing calms the nerves, relaxes the muscles and lowers the blood pressure. One interesting Korean study found that stress and anxiety were significantly reduced in a group of 60 pregnant women who practised deep breathing.[17]
2. **Notice the signs in your body**. We all store tension in our bodies. Exactly where that tension gets stored will depend on the person. For example, it could be the back, neck, shoulders or head. By better attuning ourselves to the bodily sensations associated with those areas, we can learn to respond quicker to prevent the tension developing into pain. Many people who practise mindfulness begin to notice the subtle signs and are then able to take action before it becomes a problem.
3. **Ring-fence 'me time'**. We often forget how important it is to look after ourselves, postponing the time we need to recuperate. 'Me time' is prone to becoming de-prioritised, but if it is perpetually buried under a heap of daily chores, we run the risk of burning ourselves out. Whatever you do to rest or enjoy yourself, be sure to ring-fence the time. Do not allow distractions to intrude. What are you going to do today for your 'me time'?
4. **Count your blessings – literally!** We can help banish stress by reminding ourselves of the good things we have in our lives. There is a scientifically proven benefit to doing this. Just by taking a few moments to write five things down every week can have a big impact. Those expressing gratitude this way in one psychological study reported being happier, more optimistic and physically healthier – and they actually ended up exercising more.[18]
5. **Empower yourself to take action**. People often assume that mindfulness is just about accepting the way things are. But it is also a powerful agent of change, and a way of releasing yourself from old habits. For example, at the end of a short meditation, you can set yourself a task. Instead of watching TV, this could be something as simple as getting some fresh air, calling a friend up for a quick chat, or doing some exercise.

LISTENING: THE FORGOTTEN SKILL?

Listening to others who are suffering from stress, anxiety and depression may not come naturally to us. In our schooling, we typically benefit from a much higher level of formal training in skills such as reading, writing and arithmetic. One study found that students get 12 years of formal training in writing, six to eight years in reading, one to two years in speaking, but only half a year at most in listening.[19] Yet, on average, we listen for 45 per cent of the time we spend communicating.[20] The lack of classroom emphasis on listening skills has been called the 'inverted curriculum'.[21] If we listened more attentively to each other, we would be better able to spot the symptoms of mental ill health and this would contribute to the lessening of psychological distress. Managers can model listening skills for their employees, but colleagues can also do this for each other.

Listening Out for Signs of Distress

Good listening skills can help de-stigmatise mental ill health. The tips below can help foster a better listening environment in the workplace. They are drawn from the counselling profession, and can help people who are struggling.[22,23]

- Avoid giving advice
- Sit with them in their cave
- Be a calm presence
- Learn the art of silence
- Allow anger to be heard
- Encourage dialogue on mental health.

Avoid giving advice. It may be a natural habit, but giving advice often obstructs the listening process. Less can be more here. Giving advice is largely incompatible with good listening. Although well-intentioned, it can disrupt the flow and may inhibit further dialogue.[24]

Sit with them in their cave. There is a common tendency to want to take someone out of their dark place. But attempting to cheer somebody up can be the equivalent of shutting them up if they are not in the right frame of mind. Staying attuned and focused on their words and feelings, rather than glossing over them with 'positive talk', can make all the difference.

Be a calm presence. This provides the environment for someone to trust and feel confident in opening up to you. It helps here if your body language looks engaged. Leaning forward slightly and making eye contact shows interest. There is a fine line between sounding warm and caring and sounding condescending. A genuine interest in someone else's state of mind will often pay dividends.

Learn the art of silence. In our extroverted culture, with its constant flurry of distractions, any silence is likely to be filled as quickly as it arises. Even before the smartphone era though, silences were often felt to be awkward. And where there is awkwardness, there is anxiety. But silence, properly respected and utilised, allows us the time to think and process thoughts and emotions. A conversational pause of an extra second or two creates the space for more reflection.[25,26]

Allow anger to be heard. Keep in mind that people who appear angry are likely to be feeling scared, frustrated or helpless. When they are venting in the heat of the moment, it can be incredibly hard to stay cool. The display of strong emotions in others can trigger similar reactions in ourselves. But reacting in kind is rarely the answer and can merely escalate the tension. As a rule, allowing someone a safe space to be angry without any acting out is therapeutic.

Encourage dialogue on mental health. Authentic listening is especially important when it comes to breaking down the barriers to talking about mental health.

Many workplaces appear to shy away from creating the right environment for employees to open up about how they are really feeling. The removal of hierarchical boundaries by using informal dialogue encourages open communication about mental health – and it reduces stress.

Attending Mindfully to Others

Mindfully attending to what others are saying is pivotal to good communication. Understanding what others want to communicate becomes impossible if attention is absent. People are often thinking about what they will say when a speaker finishes their turn, rather than carefully attending to what the other person is saying, and the way they are saying it. They are thinking up their arguments and how to get their points across as soon as the opportunity arises. This doesn't really qualify as communication in any real sense of the word.

If we are not concentrating on what we will say next, we are better able to pay attention to the content, meaning, tone and language used by others. Essentially, this conveys to the speaker that their contribution is of value. What is unconsciously communicated is that the speaker is respected enough to attract the undivided attention of another person.

Observing and listening to someone's verbal, vocal and bodily communication is what mindfully attending to others is all about.

THE RAILS MODEL FOR MENTAL HEALTH SUPPORT

In terms of what to do to assist a person in crisis, the RAILS model is useful in providing the right prompts to give you the confidence for handling tough mental health situations.

Remain Calm

Check in with yourself first. As a rule, you are much more able to help others if you can remain calm yourself. If a situation seems very challenging, taking a few deep breaths can make a big difference before you decide to approach someone.

Approach

Plan the best way to approach the person you are concerned about. Assess the situation as best you can. Sensitivity is required because it may be difficult for the person to open up.

Watch for signs that they may be experiencing a crisis situation:

- Alcohol or substance abuse
- Suicidal thoughts and behaviours
- Panic attacks
- Aggressive behaviours
- Trauma after an incident

- Medical emergency
- Psychotic states.

Inquire

Engage the person and ask them: “How are you feeling?” You may have noticed they are as follows:

- Behaving differently from normal
- Fatigued
- Anxious
- Stressed
- Melancholy or depressed.

Empathise with them and express your concern, but refrain from giving advice.

Listen

Listening works best if you can be non-judgemental. To do this:

- Try to put your judgements aside
- Treat the person with respect and dignity
- Keep an open mind
- Ask: “How long have you been feeling this way?”
- Give them space to tell their story.

Support

The support you provide can be practical and emotional. By being there for someone in crisis, you can:

- Give them hope for recovery
- Help them to recover faster.

Encourage the person to seek the appropriate professional support wherever appropriate. This could be in the form of:

- Workplace support
- The doctor
- Counselling or therapy.

Remember that if someone you know needs support, there are organisations that can help in difficult times:

- In the United Kingdom, the Samaritans can be reached on 116 123.

- In the United States, the National Suicide Prevention Lifeline can be reached on 1-800-273-8255.
- In Australia, the crisis support service Lifeline can be reached on 13 11 14.
- Other international suicide helplines can be found at www. befrienders.org.

THE M4 CALL TO ACTION

We all need to remain mindful of our mental health, as it can fluctuate. If you want to take up the challenge of addressing mental health issues through the use of mindfulness in the workplace, how about considering a tailored training course?

As a clinician, I can help design a programme for your organisation that makes a real difference. Many 'off-the-shelf' courses are able to raise a basic level of mental health awareness, but they are light on effective, practical coping strategies for dealing with stress, anxiety and depression. Also, they are often run by lay people who have no specialist knowledge of the field. Whilst this is not always a bad thing, these courses can sometimes pathologise mental health issues, focusing too much on 'conditions' or 'disorders' derived from an outdated medical model. They can therefore do the opposite of what they intend, stigmatising rather than de-stigmatising the subject matter.

The M4 Initiative takes a more enlightened approach, emphasising that mental health is more fluid than is often presumed, and giving people an effective set of tools to sustain a positive mindset. Integral to the course is the RAILS model for support, which is underpinned by a specific set of skills that can be taught. It is a serious subject, but one that can be tackled in an engaging way with high-quality content. Organisational commitment at all levels is essential to make it a success.

If you'd like to find out more, visit www.m4initiative.com.

KEY POINTS

- Failure to tackle mental health issues in the workplace can have catastrophic outcomes, especially for workers in safety critical roles. The Germanwings example proves that safety cannot exist without the good mental health of frontline operatives.
- Workplace stress is a key driver of mental health issues, such as anxiety and depression. Employees can find it difficult to talk about any issues they may be having out of fear of being stigmatised in the workplace. This can create a 'wall of silence' that is difficult to penetrate.
- Destigmatising the whole subject is as much of a challenge for society at large as it is for the workplace.
- Research evidence highlights how mindfulness can have a significant, long-term impact on the symptoms of stress, anxiety and depression.
- We all have innate fight-or-flight reactions. Mindfulness teaches us how to notice our reactions to stress. We can learn to perceive challenges in new ways. This enables us to respond to stressful situations with greater flexibility and freedom.

- Listening has a clear role to play in helping others in stressful situations. We can all improve our listening skills by learning from the counselling profession. This is consistent with the general approach of attending to others more mindfully.

M4: APPLYING MINDFUL SAFETY TO MENTAL HEALTH

Individual

- It is absolutely critical that we monitor our own mental health to ensure we can do our work safely.
- We can become alert to the signs of stress as they manifest themselves in our bodies.
  - For example: an increased heart rate, high blood pressure and headaches can all be symptoms of stress.
- Avoid internalising stress, if at all possible, by practising mindfulness.
- Mindfulness has specific, proven techniques for dealing with negative thoughts and feelings.
  - We do not need to change our thoughts or feelings.
  - Treating them as clouds can have therapeutic benefits.
- Appreciate the things you have to boost your mood and positivity.
- If we practise mindfulness, we can find new pathways to growth and build resilience.

Relational

- By being more compassionate and accepting towards ourselves, we can become calmer and more open to the experiences of others.
- If we notice the signs of stress in ourselves and others, we can create more harmonious social interactions in everyday life.
- By listening more attentively, we can allow difficult feelings in others to surface in an atmosphere of trust and understanding. This also enhances the quality of our relationships.
- If we can listen to the mental health stories of others, we will be in a better position to point them in the right direction to get the appropriate help.

Organisational

- How does your organisation handle mental health issues? Here are some questions worth asking:
  - Is the link between mental health and safety articulated effectively?
  - What is being done to raise awareness of mental health issues?
  - Is it possible to talk openly about mental health at work?
  - Are people stigmatised for taking time off for stress, anxiety or depression?
  - Are peer support groups available to staff in difficulty?

- Is mindfulness used as a tool to help people manage their own mental health at work?

Societal

- How does the community you live in respond to people with mental health difficulties?
- Challenge the notion that people suffering from mental ill health need to be fixed.
- Can you play a role in lessening suffering by organising a mental health initiative to help people in your community?

NOTES

1. Germanwings crash: What happened in the final 30 minutes. *BBC News*. 23 March 2017. Available from http://www.bbc.co.uk/news/world-europe-32072218 (Accessed 31.03.2018).
2. BEA (2016). *Final Report. Accident on 24 March 2015 at Prads-Haute-Bléone (Alps-de-Haute-Provence, France) to the Airbus A320-211 Registered D-AIPX Operated by Germanwings*. Available from https://www.bea.aero/uploads/tx_elydbrapports/BEA2015-0125.en-LR.pdf (Accessed 31.03.2018).
3. Ibid., p. 8.
4. Ibid., p. 87.
5. Schneidman, E. (1977). *The Definition of Suicide*. Lanham, MD: Rowman and Littlefield.
6. Ginn, S. & Horder, J. (2012). "One in four" with a mental health problem: the anatomy of a statistic. *The BMJ*. 22 February 2012. Available from https://www.bmj.com/content/344/bmj.e1302.full (Accessed 31.03.2018).
7. Gallup (2013). *State of the American Workplace. Employee Engagement Insights for U.S. Business Leaders*. Available from http://www.gallup.com/services/176708/state-american-workplace.aspx (Accessed 02.04.2018).
8. Hellebuyck, M., Nguyen, T., Halphern, M., Fritze, D. & Kennedy, J. (2017). Mind the workplace. *Mental Health America*. Available from https://www.mentalhealthamerica.net/sites/default/files/Mind%20the%20Workplace%20-%20MHA%20Workplace%20Health%20Survey%202017%20FINAL.PDF (Accessed 02.04.2018).
9. American Psychological Association (2015). Stress in America. Paying with our health. 4 February 2015. Available from https://www.apa.org/news/press/releases/stress/2014/stress-report.pdf (Accessed 02.04.2018).
10. Maestas, N., Mullen, K.J., Powell, D., von Wachter, T. & Wenger, J.B. (2015). *Working Conditions in the United States*. Results of the 2015 American working conditions survey. Available from https://www.rand.org/pubs/research_reports/RR2014.html (Accessed 31.01.2018).
11. Westfield Heath (2016). *Employees 'isolated and lost' when it comes to talking about mental health*. Available from https://www.westfieldhealth.com/press-media/press/2016/02/03/employees-isolated-and-lost-when-it-comes-to-talking-about-mental-health (Accessed 31.03.2018).
12. Kabat-Zinn, J., Massion, A.O., Kristeller, J., Peterson, L.G., Fletcher, K.E., Pbert, L., Lenderking, D.R. & Santorelli, S.F. (1992). Effectiveness of a meditation-based stress reduction program in the treatment of anxiety disorders. *American Journal of Psychiatry*, Volume 149(7), pp. 936–943.

13. Miller, J.J., Fletcher, K. & Kabat-Zinn, J. (1995). Three-year follow-up and clinical implications of a mindfulness meditation-based stress reduction intervention in the treatment of anxiety disorders. *General Hospital Psychiatry*, Volume 17(3), pp. 192–200.
14. Teasdale, J.D., Segal, Z.V, Williams, J.M., Ridgeway, V.A., Soulsby, J.M. & Lau, M.A. (2000). Prevention of relapse/recurrence in major depression by mindfulness-based cognitive therapy. *Journal of Consulting and Clinical Psychology*, Volume 68, pp. 615–623.
15. Teasdale, J.D., Segal, Z.V, Williams, J.M., Ridgeway, V.A., Soulsby, J.M. & Lau, M.A. (2000). Prevention of relapse/recurrence in major depression by mindfulness-based cognitive therapy. *Journal of Consulting and Clinical Psychology*, Volume 68, p. 616.
16. Davidson, R.J. & Begley, S. (2012). *The Emotional Life of Your Brain. How Its Unique Patterns Affect the Way You Think, Feel, and Live – and How You Can Change Them.* New York: Hudson Street Press.
17. Yu, W.J. & Song, J.E. (2010). Effects of abdominal breathing on state anxiety, stress, and tocolytic dosage for pregnant women in preterm labor. *Journal of Korean Academic Nursing*, Volume 40(3), pp. 442–452. Available from https://synapse.koreamed.org/Synapse/Data/PDFData/0006JKAN/jkan-40–442.pdf (Accessed 07.04.2018).
18. Emmons, R.A. & McCullough, M.E. (2003). Counting blessings versus burdens: an experimental investigation of gratitude and subjective well-being in daily life. *Journal of Personality and Social Psychology*, Volume 84(2), pp. 377–389.
19. Burley-Allen, M. (1982). *Listening: The Forgotten Skill.* New York: Wiley.
20. Wilt, M.E. (1950). A study of teacher awareness of listening as factor in elementary education. *Journal of Educational Research*, Volume 43(8), pp. 626–636.
21. Swanson, C.H. (1984). Their success is your success: teach them to listen. *Paper presented at the Annual Conference of the West Virginia Community College Association*, p. 23.
22. Nelson-Jones, R. (2001). *Theory and Practice of Counselling and Therapy.* London: Continuum.
23. Higdon, J. (2004). *From Counselling Skills to Counsellor: A Psychodynamic Approach.* Basingstoke: Palgrave Macmillan.
24. Howard, S. (2010). *Skills in Psychodynamic Counselling & Psychotherapy*, pp. 36–37. London: Sage Publications.
25. Ibid., pp. 46–47.
26. Milner, P. & Palmer, S. (1998). *Integrative Stress Counselling: A Humanistic Problem Focused Approach*, pp. 42–43. London: Sage Publications.

8 Culturally Mindless
The Ostrich Syndrome

Culture is the collective programming of the mind that distinguishes the members of one group or category of people from others.

– Geert Hofstede[1]

Up until now in this book, we have largely concentrated on the individual, relational and organisational levels of analysis. But the M4 approach can also be applied effectively at a cultural and societal level. To a large degree, we are mentally programmed by the culture we live in. In other words, society somehow finds its way into our heads.

This is quite similar to Freud's original notion of the superego, which contains our internalised view of our parents and society. The superego functions a bit like a societally imposed conscience, prescribing what you should and shouldn't do.[2] Attitudes, beliefs and values that exist in society at large therefore influence our own thinking and behaviour. When we are exposed to them for long enough, we assimilate them and they become part of who we are.

This is not necessarily a bad thing, of course. It often saves time and trouble if we follow societal norms. Think, for example, of the consequences that failing to observe a queue can cause in Britain. If you would prefer for a fight not to break out, it is usually better to stoically stand in line to pay for your groceries. And most of the time, following a given norm helps to preserve order and harmony in society at large. Occasionally, however, it is better to challenge the queuing norm in Britain. You could be lining up to use a cash machine and then observe a free, adjacent machine that no one else has spotted. There is no need to queue anymore in this scenario, but British minds are often set to fall into line by default, even where common sense would suggest that acting differently would be highly unlikely to offend anyone.

Sometimes, societal norms can change over a period of time. Take the Chinese habit of spitting, for example. To Western eyes, this is often viewed as a thoroughly disgusting habit. In contrast, surprising though it may seem, many people in China have traditionally viewed spitting as a cleansing habit.[3] In fact, blowing your nose into a handkerchief and then placing it in your pocket was seen as rather more disgusting – talk about spreading germs! But before the 2008 Beijing Olympics, anti-spitting campaigns were introduced in a clear acknowledgement that it might tarnish China's international reputation. Norms can be successfully challenged, especially if national pride is at stake.

In addition to national pride, lives may be at stake. In Chapter 3, the example of Air Florida Flight 90 shows that mindlessly following the societal norm of queuing can contribute to a catastrophe. Whenever people appear to be thinking the same thing, there is a real risk no one is thinking much at all. We therefore need to become aware of our own 'cultural mindset', learning to challenge it appropriately to mitigate risk. If we can acknowledge cultural blind spots, we can attain a higher degree of mindfulness and ultimately prevent accidents.

Case Study: The Meltdown at Fukushima

A form of cultural mindlessness created the conditions for the 2011 Fukushima disaster that many experts had warned of years earlier. It was a disaster that even the Japanese admitted was attributable to cultural traits – their heads had been in the sand for too long:

> What must be admitted – very painfully – is that this was a disaster 'made in Japan.' Its fundamental causes are to be found in the ingrained conventions of Japanese culture: our reflexive obedience; our reluctance to question authority; our devotion to 'sticking with the programme'; our groupism; and our insularity.
>
> – Kiyoshi Kurokawa, Chairman,
> *Fukushima Nuclear Accident Independent Investigation Commission*[4]

On 11 March 2011, a triple meltdown at the Daiichi nuclear power plant at Fukushima in Japan was triggered after a giant, earthquake-induced tsunami overwhelmed the country's north-east coast. Both the power supply and the cooling system failed at the plant after it was flooded by the 15-m high tsunami, leaving the reactor cores to melt. The tsunami went on to kill almost 19,000 people, as well as creating a truly terrifying backdrop for the world's worst nuclear accident since Chernobyl.[5] Though there were no reported deaths or cases of radiation sickness at the time, the clean-up operation at Daiichi is expected to take up to four decades at the cost of tens of billions of dollars. Nearly 7,000 workers were assigned to decommission the plant, and the reactors are still far too radioactive for humans to enter – any attempt would lead to death in a matter of minutes.

Failure by Design

Nothing short of a miracle could have prevented Japan's biggest earthquake, or the unexpectedly high tsunami that swept the north-east coast so ruinously. Nevertheless, the independent, government-mandated investigation called it a "Profoundly man-made disaster – that could and should have been foreseen and prevented".[6] The design of the nuclear plant at Fukushima had a significant role to play in the disaster.

The physical siting of the Daiichi plant was clearly unable to protect against a 15-m high tsunami. But could a tsunami that high have been predicted? Unfortunately, yes. The Daiichi plant had been built in the 1960s, at 10 m above sea level, with the sea-water pumps at 4 m above sea level. An inherent vulnerability had been locked into the plant's location by design. The original assessment for the plant's siting relied on data from the 1960 Chile earthquake, factoring in a tsunami threat of just 3.1 m.[7] The tsunami used for the siting assumption had originally started out at 25 m high, off the coast of Chile, South America. It was only once it had travelled 10,000 miles across the Pacific to Japan that it tailed off to just a few metres high. Any reasonably large, quake-induced tsunami much closer to Japan would always pose a threat to Daiichi. In the last century, there were actually eight in the region with heights at source above 10 m. On average, these occurred once every 12 years.

It was this potential danger that the Japanese ignored – and kept on ignoring. The Tokyo Electric Power Company (TEPCO), the plant operator and the Japanese regulator were locked into a cultural mindset that was immune to criticism. The International Atomic Energy Agency (IAEA) had recommended more effective provision to fight the threat of high tsunami levels as far back as 2002. This involved sealing the lower part of the nuclear facility and providing back-up for the seawater pumps, which could have bolstered the safety defences enough to protect against a tsunami.

After the disaster, both TEPCO and the regulator defended their collective inertia, claiming the height of the tsunami was unprecedented. But their argument was always weak, especially in light of the very real tsunami threat to the region. They had been warned of a potential disaster nearly a decade earlier. And each of the previous eight tsunamis in the region should have penetrated their collective consciousness.

UNFOLLOWING THE HERD

Following the cultural herd is not a trait confined solely to the Japanese. We all need to become aware of herd mentality, regardless of the culture we live in.

Stanley Milgram's 1960's obedience experiments at Yale University provide some chilling evidence of how far normal people under the influence of authority will actually go.[8] The participants, everyday folk who were from a range of occupations, were recruited from the local community and were randomly assigned to one of two groups: 'teacher' or 'learner'. The learner's role was to memorise a list of paired associates, while the teacher's role was to administer an electric shock to the learner every time their memory failed them and they gave the wrong answer to a cue word.

This happened in the presence of a white-coated 'experimenter', who proceeded to instruct the teacher to administer increasingly intense electric shocks for wrong answers given. (The shocks weren't real and the learners, perhaps better described as 'victims' here, were acting.) Yet, believing that these shocks were real, an astonishing 65 per cent of the teachers continued to administer them right to the very end of the experiment. This included potentially fatal ones of up to 450 V. They did this even though their victims had previously reported heart complaints and, in some cases, had stopped responding altogether. All that the white-coated person in authority needed to do to achieve this level of compliance from the 'teacher' was an ordered sequence of prompts ranging from "Please continue" to "You have no other choice, you must go on."

It is hard not to notice the similarity between Milgram's experiments and the blind obedience expressed by Nazis such as Adolf Eichmann for their role in murdering thousands of innocent civilians. On trial, Eichmann's defence was simply that he was obeying orders from his Nazi superiors, implying that his attitude was culturally widespread and endemic to the German population. Whether we are talking about the preference to queue, the Japanese response to a natural disaster or obeying those in authority, the capacity to step outside societal norms is undeniably important. This form of mindfulness often has a critical role to play in being able to take the best possible course of action to avert danger, human suffering or a national disaster.

But is there evidence to suggest particular national cultures may strongly influence attitudes, such as the tendencies to behave unreflectively and obey those in authority? The short answer is yes. We will need to adopt a different level of analysis to understand the impact of culture, so the next section takes a more sociological perspective.

CHARACTERISING CULTURES

This chapter started with a quote from Dutchman Geert Hofstede. Hofstede can help us understand the cultural characteristics that contributed to the disaster at Fukushima. His original 1980 work to systematically characterise different cultures within the workplace was a phenomenal tour de force.[9] He distributed a questionnaire to 117,000 managers at IBM in 40 different countries, enabling him to isolate four main dimensions:

1. **Power distance**. Would you be able to openly disagree with your manager? This dimension is all about the willingness to conform to existing hierarchies. People in societies with a high power distance are more likely to toe the line. East Asian cultures exhibit this tendency, especially in places like Malaysia and the Philippines. Conversely, people in societies with a low power distance are less likely to conform. Consequently, they feel more able to openly express their disagreements with their managers at work.
2. **Uncertainty avoidance**. Dealing with the uncertainty of life is a challenge we all face. The Swedish and Danish are typically more comfortable with this than others. They can be said to have low uncertainty avoidance. Cultures with high uncertainty avoidance, such as the Greek and Portuguese, are less relaxed about uncertainty, tending to prefer more rigid codes of behaviour. This translates into believing a company's rules or plans should not be broken.
3. **Individualism**. Do you determine your choices at work or do the group or collective usually make them? Cultures that emphasise individualism will typically allow workers more freedom to carry out a job the way they prefer. Once again, the United States, United Kingdom and Australia fit this description. Collectivism emphasises the group more, implying less personal control over tasks, with examples provided by the Latin American countries Venezuela, Guatemala and Ecuador.
4. **Masculinity**. Some cultures, such as the Japanese, heavily emphasise masculinity. This shows up in the drive for achievement, material success and asserting one's needs. On the other hand, cultures emphasising femininity promote interpersonal harmony and caring. There are some strong examples in Sweden, Denmark and Norway.

Having established that the baseline for each dimension can significantly vary across different national cultures, it is best to illustrate this point with an example.

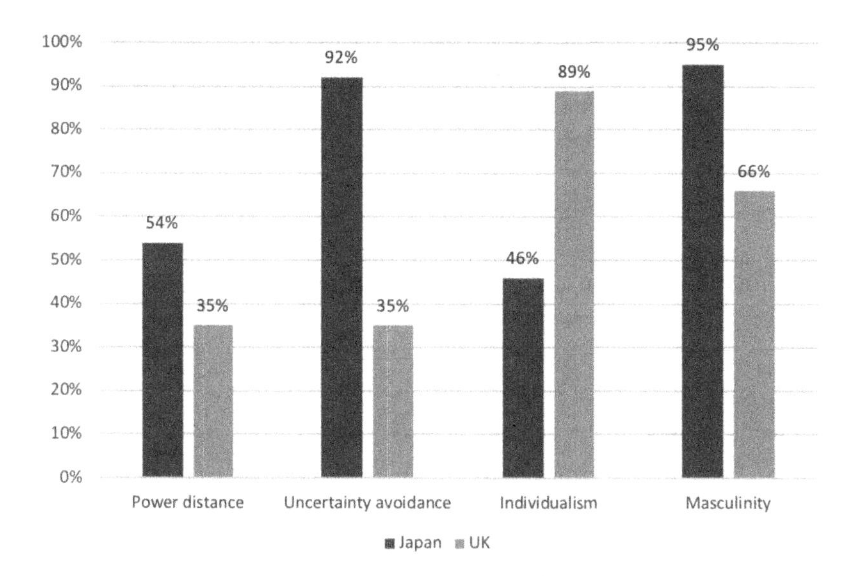

FIGURE 8.1 Japan and the United Kingdom on Hofstede's four dimensions (country comparison tool).[10]

Japan scores higher on power distance, much higher on uncertainty avoidance and much lower on individualism (see Figure 8.1).

This is supremely relevant to the national disaster that unfolded after the 2011 earthquake in Japan. The Japanese were culturally less able to challenge the status quo of existing hierarchies (high power distance), far more comfortable sticking to the accepted plan (low uncertainty avoidance) and more wedded to collective thinking (low individualism). The empirical evidence aligns perfectly with Chairman Kiyoshi Kurokawa's observations from the accident report cited earlier. The fixed, cultural mindset of the Japanese increased the safety risk and magnified the scale of the disaster.

Challenging cultural norms is bound to cause friction and confrontation, but to avoid catastrophe that is exactly what we need to do sometimes. For this to happen, we must first become aware of our cultural programming, and the cultural traits we may possess. Try entering your own country into Hofstede's online comparison tool. In terms of the results, there are, of course, no rights or wrongs. It pays, however, to reflect on how such traits may be embedded in your own way of thinking. Unthinkingly following the herd can have disastrous consequences in safety critical environments.

National Culture and Safety Culture

National culture and safety culture are closely related to each other. Safety culture refers to the norms, values and practices shared by groups, but specifically in relation to risk and safety.[11] In practice, this means that certain national characteristics will promote safe attitudes and behaviour, whilst others will not. Historically, health and safety professionals have paid relatively little attention to this relationship, but this is where organisational and societal mindfulness intersect in the M4 approach.

One European study in this exciting new field specifically looked at Hofstede's uncertainty avoidance amongst 13,600 air traffic management employees.[12] And guess what? The cultural trait the Japanese score highly on – uncertainty avoidance – was found to negatively impact safety culture.

EFFECTS OF UNCERTAINTY AVOIDANCE

We can unpack uncertainty avoidance in greater depth to reveal how this national characteristic can affect safety culture. Here are some of the potential pitfalls when the trait becomes too pronounced:

- Diverse opinions are not tolerated
- Speaking up is stifled
- Emergent risks are ignored
- Innovation is curtailed.

Diverse opinions are not tolerated. Safety management is best informed by the practice of encouraging a range of opinions. There is no 'absolute truth'. Wherever there is an assumption that there is only one way of managing safety, people become constrained by a collective mould. Fresh thinking, even if it appears 'off the wall', should always be encouraged. Dialogue on safety is hugely important – suppress it and you are likely to be heading for the next accident.[13]

Speaking up is stifled. Admitting your own mistakes, or speaking up to report something, is only really possible if managers are comfortable accepting the unknown; they must be at ease with ambiguity and uncertainty themselves. Closing one's mind to others who are bold enough to speak up shuts down a natural desire to voice concerns. Saying what's on your mind is essentially a social act that cannot happen in unconducive environments.[14]

Emergent risks are ignored. New risks emerge all the time. For example, the introduction of new staff, technology or procedures can significantly alter the operational environment. Not wanting to embrace the inherent uncertainty of a changing workplace is futile. A process of constant review is necessary to get on top of these risks and mitigate them. This enables the adjustment of strategies and the efficient reallocation of resources.

Innovation is curtailed. Old certainties can provide a false sense of security. If people gravitate towards static procedures and protocols in times of change, they will often close their minds to innovation. Complying with what we know best may be more comfortable in the short term, but it does not bode well when it comes to innovating for improved safety.[15]

Training for Extreme Discomfort

A strong preference for uncertainty avoidance can also limit the effectiveness of safety training. A study of 68 organisations in multiple industries across 14 different countries found that a high level of uncertainty avoidance typically meant employees focused on structured rather than alternative scenarios – which are the sort that require thinking on one's feet.[16] We need to train our minds to meet the challenge of uncertainty in emergency situations.

Emergencies are unstable and unpredictable by their very nature. To think on our feet and handle them effectively, we must be open-minded enough to embrace extraordinary training scenarios that stretch our thinking. We can meet the challenge of rapidly changing situations on the ground only if our minds can first be trained to work in the 'uncertainty zone' in the classroom. Naturally, this will feel uncomfortable at first. Most of us prefer to stay in our comfort zones for as long as possible. Training our minds in the uncertainty zone is a little bit like teaching our other hand to write – it is going to feel weird! With so much at stake in safety critical environments, however, it is a small price to pay.

MADE IN JAPAN: FIVE LESSONS FROM FUKUSHIMA

It may have been triggered by a massive earthquake, but even the Japanese admitted it was a disaster 'made in Japan'. Here are the main lessons we can learn from the world's worst nuclear accident since Chernobyl.

1. **Challenge national myths**. Japan created its own myth around nuclear power after the oil shocks of the 1970s. Seen as the cornerstone of energy security, nuclear power became an 'unstoppable' force. The myth became increasingly difficult to question, as it was packaged up with an exaggerated sense of national pride and self-confidence. The government bureaucracy responsible for promoting the nuclear industry also regulated it, creating a huge blind spot. Ultimately, safety was severely compromised by a nuclear myth that went largely unchallenged. Challenge such myths wherever possible, because behind them may lurk safety risks that urgently need attending to.
2. **Listen to international warnings**. TEPCO, the plant operator, described the tsunami at Fukushima as 'unprecedented'. The truth, however, is that they had been warned about the dangers many times over, but failed to act. More effective provision to fight the threat of high tsunami levels had been recommended as far back as 2002. TEPCO executives revisited the need for better coastal defences to protect against a tsunami just two and a half years before the disaster, but still no action was taken. Ignore credible safety warnings at your peril – this applies to everyone connected with safe operations.
3. **Understand local risks**. New local risks to safety may emerge over time. The Fukushima Daiichi plant had been built in the 1960s, just 10 m above sea level, with the seawater pumps a mere 4 m above it. But there were tsunamis in 1983 and 1993, with maximum heights of 14.5 and 31 m, respectively. These events should have forced a complete rethink on the safety of nuclear operations by the sea (even if they did not hit the same coastline around Fukushima). Such risks can put a massive hole in safety defences

that once seemed adequate. Pay attention to them if you do not want to get caught out.

4. **Avoid the ostrich syndrome**. A strong safety culture needs to be modelled by those in authority, but it can't be if they have their heads in the sand. The nuclear administration was reluctant to make safety an overriding priority before the accident at Fukushima. The regulator prioritised institutional wellbeing over public safety, and there was an obvious failure of the safety culture at TEPCO. Recommendations to bring safety up to international standards were habitually shelved. It is impossible to build a good safety culture if those in authority do not demonstrate it in their actions. You can help your organisation raise its head out of the sand by reporting a weak safety culture.
5. **Plan for a nationwide disaster**. No one expects a massive earthquake, tsunami and nuclear meltdown in the same afternoon, but that is exactly what they got in Japan. Imagining the worst possible scenarios is absolutely essential to having the best possible emergency preparations. Ideally, the imagination of Hollywood scriptwriters with an apocalyptic vision is needed. There was a clear failure by the authorities to plan for a complex disaster on an unprecedented national scale. Evacuation plans for the public were poorly developed and executed, creating massive confusion after the accident. If you see weaknesses in your organisation's evacuation plans, report them. Lives could depend on it.

MINDFULNESS IN SOCIETY

It is a terrifying thought that it may take a national disaster to change cultural attitudes. Japanese society was forced to reflect deeply on the cultural mindset that contributed to a nuclear meltdown, but only after the event. They were able to identify for themselves how a national emergency had been exacerbated by prevalent cultural traits. The meltdown was entirely preventable, even if the tsunami was not. There followed a sincere acknowledgment that cultural insularity had played a significant role. It is crucially important to become aware of the norms, attitudes, values and behaviours that we routinely assimilate from the culture around us. Only by reflecting upon them can we be in a position to challenge them appropriately, long before there is an accident of any kind. The example of Japan is admittedly at the extreme end of the scale, but there will be lesser examples that are no less worthy of our attention. These will play out in everyday organisational settings all over the world. Tolerance, innovation and dialogue all suffer in the absence of societal mindfulness, whilst the safety risks escalate.

Ultimately, it all depends on individual mindfulness. Cultural attitudes may permeate our thinking, but the final checks, balances, judgements and decisions reside within us. We have the cognitive ability to appraise any situation we find ourselves in and then select the optimal course of action. This may mean re-programming our brains for a different set of outcomes, but the work here is its own reward. Unshackling our minds from cultural norms promises greater freedom of choice. Fundamentally, mindfulness emphasises how we can become aware of our automatic reactions. In doing so, we create the possibility of new thinking and behaviour. It is a practice that teaches us to unfollow the herd. Often the herd resides within our own head – we have to stop it from stampeding and occasionally send it in the opposite direction.

THE M4 CALL TO ACTION

Navigating one's own values, attitudes and beliefs in organisational settings can be difficult without specialist input from outside. The M4 Initiative brings mindfulness to organisational development, allowing potentially unsafe, unhealthy patterns to surface and be examined before they threaten to cause injury, or worse, loss of life.

People are intimately connected to their organisations, their culture and their society at large. To be optimally effective, training interventions will need to emphasise both mindfulness and cultural awareness. It is possible to train people to step outside their comfort zones so they can cope better should the unexpected arise.

If you would like help in comprehensively applying the M4 approach to your unique organisational needs, visit www.m4initiative.com for more details.

KEY POINTS

- Societal norms are present in all walks of life, from everyday examples such as the queuing behaviour of the British, to the collective defence of a nation against existential threats – society gets inside our heads. In short, we are mentally programmed to a large degree by the culture we live in.
- National culture may accentuate certain tendencies, such as obedience and cultural insularity, and thus contribute to disasters that have been brewing for years. Fukushima is a strong, illustrative example.
- Hofstede's empirical research on how culture varies on different dimensions can be used to flag up where the cultural blind spots may be.
- National culture can have a strong impact on the safety regimen of an organisation.
- A cultural preference for uncertainty avoidance is strongly associated with an inadequate safety culture. Under these conditions, a lack of tolerance for diverse opinions, little desire to speak up to voice concerns, the ignoring of emergent risks and the curtailing of innovation may be present.
- We can overcome the cultural ostrich syndrome. Individual mindfulness and self-reflection are effective antidotes to ingrained, counterproductive attitudes and behaviours.

M4: APPLYING MINDFUL SAFETY TO OVERCOME THE CULTURAL OSTRICH SYNDROME

INDIVIDUAL

- An awareness of the values, attitudes and beliefs of the culture we inhabit is essential if we are to prevent disasters on a national scale.

- To prevent ourselves from mindlessly following the herd, we must challenge our inbuilt tendency to conform to prevailing norms.

Relational

- In order to prevent injuries and accidents, challenges to managers' cultural attitudes will occasionally need to be made.
- For this to occur, good employer-employee relations must be maintained through a high level of trust.
- Managers and leaders must be receptive to employee challenges, even when voiced opinions appear to conflict with deeply held cultural beliefs.

Organisational

- National culture can profoundly affect organisational safety culture. Organisations can train their employees to overcome their national culture's weaknesses and blind spots.
- By encouraging flexible thinking and responses to a range of training scenarios, health and safety risks that would otherwise go unchecked can be effectively mitigated.
- The goal is to create the right conditions for the flourishing of diverse opinions, speaking up and innovation.

Societal

- Each culture can be characterised by Hofstede's four dimensions: Power distance, uncertainty avoidance, individualism and masculinity.
- It is important to become aware of cultural blind spots, such as a high level of uncertainty avoidance, and assess how these may cascade down to the organisational, relational and individual levels.
- To avoid burying one's head in the sand like an ostrich, societies must seek international perspectives to obtain cultural objectivity.

NOTES

1. Hofstede, G. (2011). Dimensionalizing cultures: the Hofstede model in context. *Online Readings in Psychology and Culture*, Volume 2(1). Available from https://doi.org/10.9707/2307–0919.1014 (Accessed 24.04.2018).
2. Freud, S. (1978). *The Standard Edition of the Complete Psychological Works of Sigmund Freud, Volume 19 (1923–26): The Ego and the Id and Other Works.* London: Hogarth.
3. Professor studies Asian Spitting Behaviour. *BBC News*. 17 April 2013. Available from http://www.bbc.co.uk/news/uk-england-devon-22184499 (Accessed 21.04.2018).
4. The National Diet of Japan (2012). *The Official Report Fukushima Nuclear Accident Independent Investigation Commission*. The Fukushima Nuclear Accident Independent Investigation Commission. Available from https://www.nirs.org/wp-content/uploads/fukushima/naiic_report.pdf (Accessed 21.04.2018).

5. McCurry, J. (2017). Dying robots and failing hope. Fukushima clean-up falters six years after Tsunami. *The Guardian*. 9 March 2017. Available from https://www.theguardian.com/world/2017/mar/09/fukushima-nuclear-cleanup-falters-six-years-after-tsunami (Accessed 14.04.2018).
6. See note 4 above.
7. World Nuclear Association (n.d.). *Fukushima Daiichi Accident*. World Nuclear Association. Available from http://www.world-nuclear.org/info/safety-and-security/safety-of-plants/fukushima-accident/ (Accessed 14.04.2018).
8. Milgram, S. (1974). *Obedience to Authority*. London: Tavistock.
9. Hofstede, G. (1980). *Culture's Consequences: International Differences in Work-related Values*. Beverly Hills, CA: Sage.
10. Hofstede Insights (n.d.). *Country comparison tool*. Available at https://www.hofstede-insights.com/country-comparison/ (Accessed 22.04 2018).
11. Cooper, M. (2000). Towards a model of safety culture. *Safety Science*, Volume 36, pp. 111–136. Available from https://www.researchgate.net/publication/232402885_Towards_a_Model_of_Safety_Culture (Accessed 29.04.2018).
12. Noort, M.C., Reader, T.W., Shorrock, S. & Kirwan, B. (2016). The relationship between national culture and safety culture: implications for international safety culture assessments. *Journal of Occupational and Organizational Psychology*, Volume 89(3), pp. 515–538.
13. Hofstede, G. (1983). The cultural relativity of organizational practices and theories. *Journal of International Business Studies*, Volume 14, pp. 75–89.
14. Noort, M.C., Reader, T.W., Shorrock, S. & Kirwan, B. (2016). The relationship between national culture and safety culture: implications for international safety culture assessments. *Journal of Occupational and Organizational Psychology*, Volume 89(3), pp. 515–538.
15. Waarts, E. & Van Everdingen, Y. (2005). The influence of national culture on the adoption status of innovations: an empirical study of firms across Europe. *European Management Journal*, Volume 23, pp. 601–610.
16. Burke, M.J., Chan-Serafin, S., Salvador, R., Smith, A. & Sarpy, S.A. (2008). The role of national culture and organizational climate in safety training effectiveness. *European Journal of Work and Organizational Psychology*, Volume 17, pp. 133–152. Available from https://www.tandfonline.com/doi/abs/10.1080/13594320701307503 (Accessed 25.04.2018).

9 Speaking Up to Avoid Catastrophe

It takes two to speak the truth – one to speak and another to hear.

– Henry David Thoreau[1]

Mindfully speaking up not just critical for safety, but also for a healthy, happy workplace. But what happens when it falls upon deaf ears? In this chapter, we will be looking at how a failure to listen may be a reliable indicator of catastrophe. The measurable signs of a breakdown in safety critical communications are often present years before tragedy strikes.

In this context, we will be taking a closer look at the role of safety reporting. In practice, reports about health and safety can be made verbally or in writing, and they can vary in their level of formality. The situations they describe may require close monitoring, immediate action or an innovative solution in the long term. The very worst thing that can happen when a report is raised, though, is for it to be completely ignored – there isn't really a clearer way to send out a 'we're not listening' message. Not only does this discourage further reporting and feedback, but it also pours cold water over employee engagement and commitment.

SPEAKING UP

Speaking up to report things is a mindful activity that is reliant, in part, on observing what is going on around us externally. But it is also about verbalising the stuff in our heads, so when we notice that we do not feel safe we are able to summon up the courage to find the words. Fully enabling the ability to speak up is a massive challenge and it is dependent on a range of factors across all four levels within the M4 approach. Even where workplaces foster an open, trusting atmosphere, employees may struggle to voice their concerns. And it is equally true to say that concerns articulated by employees run the risk of being routinely ignored. Before change can happen, there must be a meeting of minds. For every health and safety concern raised by an employee, there must be someone in authority willing to listen.

If we hold a position of authority, a mindful approach can help us accept any perceived negatives reported, collect our thoughts, investigate with an open mind, and then respond with greater freedom and flexibility. When initially confronted with a difficulty or challenge, we usually have more options than we imagine. Instead of adopting a dismissive stance towards people who voice concerns, perhaps even labelling them as 'troublemakers', we can usefully ask ourselves what we can learn from the feedback. Whatever the content or nature of the report itself, we can always treat it as an opportunity to learn about the effectiveness of reporting mechanisms within an organisation. This means looking beyond our personal reactions to the report to appreciate the bigger picture.

Case Study: The Failure to Listen at Grenfell Tower

A blackened shell was all that remained of the tower that witnesses said, "Went up a like a matchstick".[2] The fire that broke out shortly after midnight on 14 June 2017 on the fourth floor of the 24-storey Grenfell Tower started in a fridge freezer. It went on to claim the lives of 72 people, whilst leaving hundreds homeless in West London.[3]. The external cladding material used in a recent refurbishment spread rapidly to engulf the 127 flats in the building; it contained polyethylene, a material so flammable it has been likened to petrol.

Residents had raised their safety concerns four years prior to the tower block fire. As far back as 4 February 2013, The Grenfell Action Group had warned that fire safety equipment had not been tested for 12 months.[4] Minutes from an emergency residents' meeting in March 2015 detailed a long list of concerns about the refurbishment work, citing "cheap materials and corner cutting".[5] The litany of serious health and safety failings also included flammable parts to the window frames, a smoke extraction system unfit for purpose and flawed building regulations, allowing systems to introduce obvious dangers from the design stage onwards.[6]

The Grenfell Action Group's chillingly accurate warnings reached fever pitch in a blog posted in November 2016, just months before the tragedy.

> It is a truly terrifying thought but the Grenfell Action Group firmly believe that only a catastrophic event will expose the ineptitude and incompetence of our landlord, the KCTMO [Kensington and Chelsea Tenant Management Organisation], and bring an end to the dangerous living conditions and neglect of health and safety legislation that they inflict upon their tenants and leaseholders.[7]

The residents' prediction of catastrophe correlated well with prior international experience of tower block blazes. Before Grenfell, there were already tragic examples of flammable cladding accelerating fires in other countries. Australia had a high-rise fire in 2014 and subsequently imposed a ban on unsafe cladding.[8] Dubai had one in 2015 and similarly used a ban on high-risk materials to mitigate the risks.[9] But the lessons that could so easily have been learned from overseas were mindlessly ignored. No one was listening.

A tragedy on the scale of Grenfell will make headline news for weeks, but the narrative of inaction behind it showed how health and safety risks had been ignored for years. The main risks had been identified, but residents' concerns were repeatedly met with a wall of indifference. Grenfell is an example of so many things going wrong in slow motion, culminating in the worst possible outcome: many lives pointlessly destroyed in a preventable fire, with hundreds of others wrecked.

In the case of Grenfell, residents found their voice and spoke out long before catastrophe struck, but the authorities and decision-makers did not listen to their valid health and safety concerns. A similar pattern can also be found in organisational settings, where there are many barriers that can potentially thwart timely action in relation to health and safety reports. Besides helping avoid catastrophe, the intelligence provided by health and safety reports prove invaluable for organisational learning. Before we take a closer at how we can overcome some of these barriers, it is worth examining how something called the 'negativity bias' can impact upon and potentially stifle the desire to report.

THE NEGATIVITY BIAS AND FEAR

The negativity bias refers to the notion that harmful or traumatic events, as well as unpleasant thoughts or emotions, have a disproportionately strong effect on our state of mind, in comparison to positive or neutral ones. It is a useful notion in the context of reporting, where we are usually concerned with safety incidents of a negative nature. For many people, this will often bring to mind events that can cause fatalities or serious injury – these can understandably trigger a fearful response.[10]

In this context, *negative events* can refer both to those which have occurred in the past, or those envisaged in the future. Possible future events may be perceived as threatening and potentially harmful in much the same way that past events do.[11] Consider what goes through a potential reporter's mind when weighing up whether to speak up or not, either about an incident or unsafe practice. The negativity bias is likely to come into play. Whether the incident has caused physical or mental harm, or simply threatened to cause harm, the effect may be just the same.

The brain perceives both scenarios as threats of a similar magnitude. In response to the anxiety induced, an attempt may even be made to deny the negative event altogether.[12] If the event is denied, or avoided because it is highly uncomfortable, it will never make it into a report of any kind. It is therefore crucially important to put the onus on creating an environment which feels psychologically safe enough for someone to instigate a report about an adverse event.

OVERCOMING THE BARRIERS TO SPEAKING UP

Let's take a look at the healthcare environment, where some important research can fruitfully highlight some of the key obstacles to speaking up. Take childbirth: delivering babies routinely throws up all sorts of reportable issues. Anaesthetic problems, blood transfusions and occasional deaths all occur in such environments, but they aren't necessarily reported. This is why Charles Vincent, a psychologist and expert in patient safety, chose two obstetric wards for his research involving 156 obstetricians and 42 midwives.[13] These busy wards were ideal for exploring the reasons why staff did not always speak up, but also for finding practical ways of remedying this.

Knowing What Is Reportable

If staff members do not know what to report in the first place, or it is left rather vague, they are far less likely to do so. During his research, Vincent found that 30 per cent of staff did not know where to find a list of reportable incidents in the obstetric wards, even though most staff knew that a reporting system existed. Workers need to understand what constitutes a reportable incident. It begs the question of how staff can be expected to report at all, if they do not know what kinds of events they should be reporting.

Ninety-six per cent of staff said they would always report a maternal death. What is most striking, perhaps, is that a greater percentage of staff said they would always report a major complaint compared to a stillbirth/neonatal birth (see Table 9.1). It is difficult to fathom how a major complaint could rank higher for reporting than a stillbirth/neonatal death. Or looking at it from another angle, 14 per cent of staff would not always report a stillbirth/neonatal death, despite loss of life being involved.

For less serious outcomes, there was a proportionate decline in the number of staff who said they would always report. But there may be a wealth of data in adverse events that are not ranked as highly, such as anaesthetic problems, extended third-degree tears or blood transfusions. If we have less data on these events because fewer staff will commit to always reporting them, the opportunity to learn from such events is left untapped.

This all highlights that if we want to build a full picture of all the adverse events that take place, and avoid losing significant data, we must draw up a list of all the reportable adverse events. The sharing of lists can boost reporting in any industry where metrics are required and provide valuable intelligence. The same technique could also be applied just as straightforwardly in industries beyond healthcare, such as energy, rail and aviation.

TABLE 9.1

Staff in Obstetric Wards Who Will Always Report, by Incident Type

Incident Type	Would Always Report (Per cent)
Maternal death	96
Major complaint	88
Stillbirth/neonatal death	86
Convulsions	73
Anaesthetic problem	66
Extended third degree	55
Unexpected admission to special care baby unit	39
Blood transfusion	20

Source: Adapted from Vincent et al (1999).

Ignoring Less Serious Events

We also need to counteract the bias in favour of always reporting more serious events, potentially at the expense of reporting the lesser ones. To address this, the reporting of less serious events should be encouraged with positive feedback. It should also be emphasised that unpacking the nature of less serious events can yield important intelligence about the kind of organisational culture present.

For example, an inadequate pre-operative assessment by an anaesthetist could later contribute to complications for a patient in surgery, such as vomiting or nausea. If the patient recovered quickly, it might seem less important to report the incident. However, further investigation might reveal the assessment had been rushed because of a high workload. This 'root cause' could have far more serious consequences for a patient on another occasion, and not just in anaesthetic practice. Therefore, it is important to determine the cause so that management can address any systemic issues.

Lists needn't just include adverse events with potentially severe consequences. The ideal system will include in its design the positive reporting of events too – something that is frequently overlooked. We will explore this in greater depth later in the next chapter.

Why Report If It Ended Well?

Where employees feel that they have dealt with a situation effectively to avert a negative outcome, there is less motivation to report. As in the anaesthetic example above, what incentive is there to report if they have recovered the situation and no one got hurt? Making a report may just seem like unnecessary effort. There is evidence to suggest that unusual anaesthetic incidents are almost always reported, but less than 20 per cent of the common ones are left unrecorded. This is primarily because possible harm has been successfully averted.[14]

There may be little incentive to report potentially negative outcomes that have been transformed into positive ones, but this represents a massive, untapped learning opportunity. We can learn at least as much from the things that go right as from the things that go wrong, and probably a lot more. The skewed emphasis on negative outcomes comes from constant reinforcement by regulators, authorities and organisations that are almost exclusively interested in learning from incidents.

But as Hollnagel points out, whether we are talking about a patient admitted to an emergency room, or a car journey, the probability of failure is roughly 1 in 10,000.[15] This leaves the other 9,999 events where things go right; for example, in the form of a successful medical procedure, or an accident-free car journey. These events are usually ignored, and the learning potential from them remains under-utilised.[16] In short, this creates a grossly asymmetrical learning situation where organisations often prefer to learn from failure rather than from success. Most learner drivers would be horrified to hear that their instructor favoured analysing car crashes to help them pass their driving test quicker! Yet this is analogous to the approach predominantly taken in health and safety management.

Top Five Reasons for Not Reporting

The findings from Charles Vincent's study of reporting in a healthcare environment can be applied more broadly to other high-hazard industries where safety is of paramount importance. Blame was the top reason staff cited for not reporting.[17] A huge subject in its own right, it is counterproductive, overplays the role of individuals and frequently downplays organisational factors. For this reason, we will be tackling the subject of how we can proactively turn blame into trust separately in Chapter 10. Organisations will also need to have a plan of action for tackling some of the other main reasons listed below.

1. **Blame**. Apart from anything else, it never creates safer environments.
2. **Fear**. Often generated by a blame culture, it kills off the very thought of reporting.
3. **Lack of time**. The busier you are, the less likely you will be to report something.
4. **Apathy**. If employees feel that no one is listening, they'll soon stop reporting, no matter how much you encourage them.
5. **A negative view of reporting**. This isn't surprising, given that our views are normally coloured by high profile incidents with nasty repercussions.

DESIGNING AN EFFECTIVE SAFETY REPORTING SYSTEM

If we want to capture all the useful safety intelligence we possibly can and prevent incidents, the tips below will help design a reporting system that works effectively.

Lead by example. If a reporting system is to work optimally, employees in senior positions need to set the benchmark high and lead by example. In Charles Vincent's study, midwives said they were more likely to report incidents than doctors, and junior staff were more likely to report than senior staff.[18] Senior doctors could no doubt point to the length of their 'to do' list as an excuse, but you cannot imagine junior midwives twiddling their thumbs during the delivery of babies either! Everyone must play their part.

The more eyes the better too. If all grades of staff can report, it sends out a message that an organisation is inclusive and values contributions to health and safety culture from all quarters.

Make reporting easy. To counteract the main reasons for not reporting in the first place, the process must be made as simple as possible in order to ensure it does not become burdensome, or an afterthought. Mobile applications have gained popularity in recent years, and they can relieve the perceived burden of reporting. They can make it easier to speak up, sometimes providing staff with an option to report anonymously should they be worried about being blamed.

Anonymous reporting, though, means that the reporter cannot be contacted to provide further information. If the quality of information provided by an anonymous reporter is dubious, their information may be difficult to verify or substantiate. Some researchers have also pointed out that staff may still not report, even with the provision of an anonymous route.[19]

Close out every single report. If we want people to report, we need to be in the habit of providing feedback. Without it, people soon grow tired of making the effort. Reports must not end up in a bureaucratic black hole. Providing a response every single time sends a clear message that an organisation is listening. Implementing such an approach may increase the volume of reports by ten times, vastly increasing the storehouse of actionable intelligence.[20]

Use confidential reporting for back-up. No in-house reporting system will be able to detect and capture every health and safety concern with complete certainty. Confidential reporting is there to stop concerns slipping through the net.

CONFIDENTIAL REPORTING TO REDUCE FEAR

Fear has the clear potential to kill off health and safety reports altogether. On one level, it stops people reporting in the first place; on another, it makes them worry about the repercussions if they do come forward. In such situations, critical information that could prevent an accident may not surface until it is too late.

Complimenting existing reporting channels, confidential reporting provides another avenue for employees to raise concerns, but with the assurance that their identity will not be revealed. In the interest of establishing an emotionally safe place for staff to discuss their concerns, confidential reporting is best provided by an independent body. With the prevention of safety incidents in mind, confidential reporting schemes often focus on precursor data provided by near misses or close calls. They are distinct from 'whistleblowing' services that may eventually result in an individual being identified in a court of law.

Typical features of a confidential reporting scheme include:

- Independent, third-party provision of a reporting channel to all employees.
- Special processes to ensure no confidentiality breaches.
- The facilitation of difficult-to-resolve health and safety issues.
- A newsletter to share the learning from confidential reports.

Organisations attempting to provide their own confidential reporting schemes are likely to find that their employees have difficulty trusting them. Sometimes, the organisation is at fault, showing an inclination to dismiss valid health and safety concerns. Viewing concerns through a culturally biased, organisational lens potentially misses important areas of risk. This is a form of 'groupthink', where critical evaluation can suffer at the expense of maintaining group harmony and the status quo. Symptoms can include a tendency to ignore or discredit information contrary to a group's position, and the use of direct pressure to bring dissidents back into line.[21]

Interestingly, there are only a few confidential reporting schemes worldwide that meet the criterion of being independently operated. This is rather surprising given their potential to uncover and address hidden risks that affect the bottom line. In the United Kingdom, independently operated, confidential reporting schemes include CIRAS (Confidential Incident Reporting and Analysis Service) for transport workers, and CHIRP (Confidential Reporting Programme for Aviation and Maritime). In the United States, C^3RS (Confidential Close Call Reporting System) covers selected railroad carriers.

Confidential versus Anonymous Reporting

People often confuse anonymous and confidential reporting, but there are fundamental differences between them. With anonymous reporting, the identity of the reporter is not known, because it is not disclosed. This creates two main difficulties. Firstly, it means the accuracy of the information may be impossible to verify, where, for instance, you cannot ask follow-up questions on the facts or enquire about motivations for reporting. Secondly, it is not possible to provide feedback to the reporter, who may be able to comment on any improvements on the ground.

With confidential reporting, the identity of the reporter is retained by the scheme's operator, but no one else, for the duration of the reporting life cycle, before being erased from the system. This allows for the full verification of the information supplied, whilst closing the feedback loop by providing a response to the reporter and seeking their comments. This means the issues raised can be resolved to the satisfaction of all the parties involved.

KEY POINTS

In order to play an influential role in organisational learning, health and safety reporting ideally needs to:

- Encourage high rates of reporting by emphasising its importance.
- Emphasise the need to report big and small incidents.
- Create a list of reportable incidents and publicise these.
- Recognise the value of reporting in recovery situations.
- Tackle blame, fear, lack of time, apathy and negative perceptions of reporting.
- Ensure everyone has a commitment to reporting, regardless of seniority level.
- Make reporting as easy as possible – for example, by using mobile applications.

M4: APPLYING MINDFUL SAFETY TO IMPROVE SAFETY REPORTING

Individual

- To prevent tragedies like Grenfell Tower, we all have a role to play in mindfully noticing things that are potentially unsafe.
- Once we've noticed something unsafe, it takes courage to speak up and report our safety concerns because it can feel like the brain is under threat.
- We need to make time to report any concerns and overcome any sense of apathy.
- It is important to report less serious events, and events that ended well (though could have had nasty consequences), as they can yield important insights.
- Finding the courage to speak up and report can save lives. If we can't report using conventional channels, reporting confidentially may provide a viable option to surface hidden intelligence.

Relational

- Upon receiving a safety report of any kind, it helps to adopt a responsive, listening approach.
- Even if we don't agree with the content of a report, it may tell us something useful about the reporting system itself – or how the wider organisation processes feedback.
- Managers will need to actively encourage safety reports from their staff to ensure safety critical information is captured as completely as possible.

Organisational

- The promotion of speaking up and reporting safety issues is an integral part of a mindful safety culture.
- Senior management support for a robust, easy to use reporting system is of crucial importance to ensure effective learning from incidents.
- All grades of staff should be encouraged to report for a 'maximum number of eyes on ground' approach.
- Independent, confidential reporting can be used as a back-up option to ensure safety reports don't slip through the net.

Societal

- Responsibility for safety can stretch well beyond organisational settings into society, involving many actors – for example, private citizens, landlords, councils and private companies (as in the case of Grenfell Tower).
- Consequently, speaking up to report safety concerns can additionally be construed as a collective responsibility held by groups or whole communities.
- In society at large, we need to remain vigilant, and feel comfortable reporting safety concerns, because failing to act can detrimentally affect us all.

NOTES

1. Thoreau, H.D. (2000). *A Week on the Concord and Merrimack Rivers*. London: Penguin Books.
2. Mostrous, A. & Goddard, L. (2017). Witnesses say panels on outside of Grenfell Tower 'went up like matchstick'. *The Times*. 15 June 2017. Available from https://www.thetimes.co.uk/article/witnesses-say-panels-on-outside-of-flats-went-up-like-matchstick-s2pflnxvp (Accessed 31.05.2018).
3. What happened at Grenfell? *BBC News*. 18 May 2018. Available from http://www.bbc.co.uk/news/uk-england-london-40272168 (Accessed 31.05.2018).
4. Concerns raised about Grenfell Tower 'for years'. *BBC News*. 14 June 2017. Available from http://www.bbc.co.uk/news/uk-england-london-40271723 (Accessed 31.05.2018).
5. Lusher, A. (2017). Grenfell Tower residents complained two years ago of 'cheap materials and corner cutting' in block's refurbishment. *The Independent*. 16 June 2017. Available from https://www.independent.co.uk/news/uk/home-news/grenfell-tower-fire-residents-fears-warnings-ignored-kensington-chelsea-borough-council-tmo-a7794086.html (Accessed 31.05.2018).

6. Booth, R., Bowcott, O. & Davies, C. (2018). Expert lists litany of serious safety breaches at Grenfell Tower. *The Guardian*. 4 June 2018. Available from https://www.theguardian.com/uk-news/2018/jun/04/expert-lists-litany-of-serious-safety-breaches-at-grenfell-tower (Accessed 07.06.2018).
7. Grenfell Action Group (2016). *KCTMO – Playing with fire!* Blog. Available from https://grenfellactiongroup.wordpress.com/2016/11/20/kctmo-playing-with-fire/ (Accessed 31.05.2018).
8. Oaten, J. (2019). Builders on notice: don't use flammable cladding on Victorian high-rises. *ABC News*. 10 March 2018. Available from http://www.abc.net.au/news/2018-03-10/victorian-government-moves-to-restrict-use-of-flammable-cladding/9534926 (Accessed 31.05.2018).
9. Dubai bans use of flammable materials in buildings. *Emirates 24/7 News*. 21 April 2016. Available from https://www.emirates247.com/news/emirates/dubai-bans-use-of-flammable-materials-in-buildings-2016-04-21-1.627946 (Accessed 31.05.2018).
10. Rozin, P. & Royzman, E.B. (2001). Negativity bias, negativity dominance, and contagion. *Personality and Social Psychology Review*, Volume 5(4), pp. 296–320.
11. Taylor, S.E. (1991). Asymmetrical effects of positive and negative events: the mobilization-minimization hypothesis. *Psychological Bulletin*, Volume 110(1), pp. 67–85.
12. Kubler-Ross, E. (1969). *On Death and Dying*. New York: Macmillan.
13. Vincent, C., Stanhope, N. & Crowley-Murphy, M. (1999). Reasons for not reporting adverse incidents. *Journal of Evaluation in Clinical Practice*, Volume 5(1), pp. 13–21.
14. Jayasuriya, J.P. & Anandaciva, S. (1995). Compliance with an incident reports scheme in anaesthesia. *Anaesthesia*, Volume 50, pp. 846–849.
15. Hollnagel, E. (2014). *Safety-I and Safety-II. The Past and Future of Safety Management*. Ashgate: Farnham.
16. Ibid.
17. Vincent, C., Stanhope, N. & Crowley-Murphy, M. (1999). Reasons for not reporting adverse incidents. *Journal of Evaluation in Clinical Practice*, Volume 5(1), p. 17.
18. Ibid., pp. 16–17.
19. Hart, G.K., Baldwin L., Gutteridge, G. & Ford, J. (1994). Adverse incident reporting in intensive care. *Anaesthetic Intensive Care*, Volume 22, pp. 556–561.
20. Kaplan, H.S. & Fastman, B.R. (2003). Organization of event reporting data for sense making and system improvement. *Quality and Safety in Health Care*, Volume 12(II Suppl), pp. ii68–ii72.
21. Janis, I.L. & Mann, L. (1977). *Decision Making*. New York: Free Press.

10 Mindfully Learning from Positives

Many good events can overcome the psychological effects of a single bad one.

– Roger F. Baumeister[1]

As we saw in the last chapter, health and safety reporting is normally associated with adverse events. This is part of the reason why it is so difficult to get people to speak up and report in the first place. Adverse events naturally make us clam up. They may bring back bad memories or unwittingly lead us to relive the trauma of an incident that possibly resulted in serious injury, or even death. They are often associated with blame, disciplinary hearings or litigation.

We also discussed the 'negativity bias' and lots of practical tips for reducing the psychological barriers to speaking up in this context. In this chapter, there is a fundamental change in emphasis towards the learning that can be obtained from the largely untapped reservoir of *positive* events. In fact, the whole notion of what should be reported needs to shift for these barriers to be overcome. This doesn't imply in any way that we should stop recording adverse events – only that we need to redress the imbalance in favour of far more positives.

Making the shift in thinking from negative to positive reporting is likely to require a quantum leap for most of us – and with good reason. Our brains are wired for fear because they are the product of millions of years of evolution. This means we are going to need to commandeer all our mental resources to effect the change we wish to see towards positive reporting. The effort to report positive events may not feel entirely natural, but this is no reason not to try.

REWIRING OUR BRAINS FOR POSITIVITY

Our ancestors had physical survival uppermost in their minds. They were constantly scanning their surroundings for the next threat to life or limb. Their brains naturally had to be ready for 'fight or flight'. It is not surprising, therefore, that negative events started to have more of an impact than positive ones. Once bitten, twice shy, as the saying goes. All of this can make us very afraid of reporting, because our brains are predisposed to fear negative events. Drawn to bad news, we find it far easier to store negative events and, even when the threat is no longer there, our minds have an irritating habit of calling them up.

Neuroscientist Rick Hanson says the brain is like Velcro for negative experiences, but like Teflon for positive ones.[2] This means that we must work much harder to make positive experiences 'stick' because they are quickly forgotten otherwise. On the other hand, negative experiences are far more difficult to erase from our thinking and memories.

THE GOLDEN POSITIVE TO NEGATIVE RATIO

There's so much media-induced negativity around and we are subjected to a relentless stream of stories and images 24/7, 365 days of the year. Despite this, it is possible to design safety reporting systems which put much greater emphasis on positive events to maximise organisational learning. Focusing more on positive events also helps normalise the practice of reporting, encouraging the habit whilst making the whole experience far more enjoyable. Somewhat ironically, the experience of reporting positive events and receiving encouraging feedback increases the chances that adverse events will be reported more too.

This all begs the question of what ratio of positive to negative events to aim for when we apply the same thinking to health and safety reporting in organisational settings. Rather than arbitrarily plucking a figure out of the air, we can take our cue from some fascinating research. In the sphere of close personal relationships, it typically takes about five positive interactions to overcome the effects of a single negative one. A less favourable ratio is a predictor of marital discord and the slide into eventual break-up.[3]

In the search for the 'Ideal Praise-to-Criticism Ratio' in organisations, top performing teams were found to give an average of 5.6 positive comments for every negative one.[4] It was the factor that made the biggest difference between the most and least successful teams. We can apply this finding to the world of reporting and perhaps even aim a little higher. Six positive events reported for every negative one is an achievable target. The reporting infrastructure will need to be in place to accommodate this change in thinking. Reporting forms should nudge staff into reporting positive events more, whilst incident databases will often have to be redesigned to follow suit. A completed reporting card designed with this in mind might look something like the one presented in Figure 10.1.

GOOD RELATIONSHIPS AND SAFETY

The working relationships managers have with their employees are critical for allowing safety reports to surface. To overcome some of the ingrained biases that stop us from speaking up, the shift to positive reporting can make a huge difference. This doesn't mean ignoring the negatives – on the contrary, it means creating an environment where people feel comfortable to report those too.

We are aiming for proportionately more positives than negatives, but somewhat ironically, this will likely lead to a greater number of negatives being reported overall. Any approach exclusively emphasising the reporting of negative events will fail to optimise their capture. And there is a huge benefit to increasing the reporting of positives – the whole process enhances relationships.

Consider the C³RS reporting intervention at one site that saw a statistically significant 41 per cent reduction in derailments. In this case, confidential reporting was used as a tool to restore trust. The intervention also helped reduce the de-certification of locomotive engineers – where authorisation to operate trains was removed – by 31 per cent.[5] These reductions were achieved in the context of improved safety culture

Golden Ratio reporting card		
If you are happy to provide contact details, we will be able to get back to you with a response: ***Name:*** *Joseph Stone* ***Phone:*** *07906 795 345* ***Email:*** *jstone@reportingheaven.com*		
I would like to report...		
Something going wrong:	***Things going right:***	
I wasn't given my safety briefing today. I felt uncomfortable starting work this way.	1.	*My supervisor usually listens to my concerns.*
	2.	*I feel able to speak up and report safety issues.*
	3.	*Yesterday, the worksite felt very safe. Knew what to do.*
What I can do to change the situation:	4.	*My colleagues were supportive today - asked for help, got it.*
Report it here and ask for feedback. Bring it to my supervisor's attention.	5	*We finished on time today. This was down to good planning.*
	6.	*I feel like I've done a good job. I'm going home satisfied.*

FIGURE 10.1 Example of a 'golden ratio' reporting card.

and employee engagement. As one interviewee put it, "Filling out a C³RS form makes you think about what happened, so you are less likely to do it again." Essentially, the act of reporting makes one more reflective and mindful.

Relationships benefitted at many different levels. Collaboration was visible right the way through the intervention. A peer review team comprising frontline staff, management and the Federal Railroad Administration undertook root cause analysis and problem solving. Senior managers also monitored corrective actions in response to close call events. The results of the changes were then communicated back to employees to close the loop.

Interviewees said that the relationships between frontline staff and managers improved, with senior managers expressing their satisfaction with this enhanced employee engagement. The pursuit of a reduction in safety incidents aligned with better risk awareness and improved vertical and lateral communication – it paid financial dividends, too.[6]

Positive Safety Scripts

The C³RS example shows that in response to a problematic situation, it is possible to create a new, positive script for both workplace relationships and safety. This is best achieved through close collaboration and shared safety goals – these are all key elements of a mindful safety culture.

Taking this one step further, the same principles can be applied to situations likely to pose challenging safety risks in the future. This is all about adopting a preventative mindset to visualise the safety of people and other assets, then implementing a plan to ensure this happens. A positive, future oriented safety script will therefore:

- Visualise positive safety outcomes for all those involved in operations.
- Carefully assess all the known risks by fully brainstorming them.
- Consider a variety of possible emergencies and 'dreamed up' scenarios to encourage flexible thinking for dealing with hidden or currently unknown risks.
- Train or 'dress rehearse' these scenarios to test operational readiness.

The case study of Prince Harry and Markle's wedding shows how this can be achieved in practice.

Case Study: The Royal Wedding

Prince Harry and Meghan Markle's wedding actually provides a great example of how we can positively orchestrate the identification and control of substantial health and safety risks, with the goal of averting an unthinkable tragedy. The media is constantly able to attract our attention by playing on our worst fears, but the normal rules of the game were dutifully suspended in the case of the royal wedding. More than two billion people worldwide watched the event, and a security operation costing an estimated £30 million was deployed to keep everyone safe, whilst ensuring the spectacle unfolded as seamlessly as possible before the cameras.[7,8]

The town of Windsor, with its narrow, cobbled, publicly accessible streets, was referred to by the former Head of Royal Protection as a "hell of a job to secure".[9] The event was a prime target for a motley crew of terrorists, protesters, royal stalkers and wayward drunks. Months of planning, the systematic identification of hazards, the control of known risks and the anticipation of possible emergencies proved indispensable for successfully managing the vast crowd of around 100,000 on the day. No bombs went off, no drunks or stalkers got near the royals and, more to the point, virtually everyone enjoyed the celebration in the May sunshine. 'Elf and safety' created the perfect conditions for the royal fun, but I do not believe any news outlet ran that particular headline, suggesting old media habits die hard.

Controlling the Risks to the Royals

The measures implemented prevented a major incident through the clear identification and control of risks. Police and Special Forces had to be ready for terror attacks that could have involved knives, vehicles or suicide bombers.[10] The entire route of the royal procession had been walked and assessed months in advance, with the layout of Windsor Castle closely studied. Security prevailed on the day through the use of hi-tech resources and a strong police presence.[11] Every angle was covered.

- **In the air**. Helicopters patrolled the skies as a constant high-profile reminder that everyone was being watched. A no-fly zone was implemented and a defence system was deployed to jam any drones attempting to enter the area.
- **On the river**. Highly visible, marine police units were deployed on the River Thames.
- **On the roads**. Pinch-points that could help would-be attackers were identified in advance, whilst escape routes and safe locations were determined just in case. Road closures were implemented to thwart terror attacks on large groups of pedestrians and hostile vehicle barriers and automatic number plate recognition technology were used. Areas of high risk for when the royal carriage drove through Windsor were monitored.
- **On the ground**. The presence of armed police, plain-clothes officers, sniffer dogs to detect explosives, airport-style security checks and 'stop and search' checks reduced the security risks. There was also a 'ring of steel' perimeter around Windsor Castle itself. Bins were replaced with plastic ones that couldn't be moved, preventing their potential use as weapons.
- **On rooftops**. Snipers had the advantage of seeing events – and any potential threats – from their unique vantage point.
- **Stalking the stalkers**. Police viewed photos of obsessed stalkers in advance of the wedding. The plan involved surrounding a stalker with plain-clothes officers who would then discreetly usher him or her away. Even before the wedding, known royal stalkers were visited to assess the risk they posed.[12]
- **Dress rehearsal**. Part of a former RAF base was transformed into a replica of Windsor High Street to simulate the conditions on the day. A dry run was performed with a range of staged events, from anti-monarchy demonstrations to full-blown attacks. An Iraq veteran and his wife played the royal couple and they were assailed by all manner of unthinkable horrors that included chemical attacks, snatch operations by would-be kidnappers and even a grenade attack that forced the pair into a ballistic blanket for safety. To prevent a terrible, alternative wedding, worst-case scenarios were brought to life with the skill of a Hollywood movie director.

All the security personnel involved were galvanised by the goal of ensuring the safety of the public and the 80 members of the royal family in attendance. They dreamt up surreal-sounding scenarios to ensure the 'impossible' did not become reality and destroy the celebrations. The hard work paid off. There were no significant incidents of note on the day.

LEARNING FROM POSITIVE EXPERIENCE

In the last chapter, we mentioned how learning from positive events, where things have gone right, usually represents a big, untapped learning opportunity. If we can effectively translate a positive safety experience from one setting to another, there is a strong possibility it may be able to produce a good safety outcome in the new setting too.

This shifts the emphasis to what we can do, rather than what we can't, to ensure successful safety outcomes. It stops us being bogged down in problems, steering us towards potential solutions instead. Our frame of reference correspondingly shifts to pinning down what works effectively, thus opening up a new, productive line of appreciative enquiry. It also informs the kinds of questions we can ask in pursuit of optimal safety performance.

For example, rather than exclusively focusing on what distracts us from a task such as driving, we can also ask how we can maintain a high level of alertness behind the wheel. Or rather than merely pinpoint where safety communications are possibly failing, we can shift our thinking to ask what skills are needed for mindful communications. Once we identify these positive examples, they can serve as exemplars for others to follow. We can extend the same principles to whole organisations, societies, and national governments.

It can even work for one of the greatest health and safety challenges of modern times: the Covid-19 pandemic. Though the potential for a pandemic of this nature was an acknowledged possibility, its rapid spread took people by surprise. Even Dr Anthony Fauci, the US's top infectious disease expert, hadn't realised how rapidly it would take over the planet.[13] In just four months, seven million had been infected and over 400,000 had died. By all accounts, the contagiousness of the virus had been grossly underestimated.

Despite the shell shock, however, some countries fared much better in their responses to the shared public safety challenge. An instructive example is provided by the response in Hong Kong care homes.

Case Study: Resilience in Hong Kong Care Homes

When Coronavirus was ravaging care homes across Europe and America and killing tens of thousands, Hong Kong was able to highlight a very different experience – not a single care home resident even contracted Covid-19.

Hong Kong's apparent success was down to several important factors. Moreover, its resilient approach was able to provide a model for others to follow.

Rebooting the collective memory. Back in 2003, Hong Kong became the epicentre of the SARS outbreak with 299 deaths. Just as with Covid-19, the elderly were the most susceptible to the virus. Two nursing home workers died, and 54 nursing homes ended up with cases. The psychological scars were still present 17 years later when Covid-19 struck, but the traumatic memories ensured Hong Kong's nursing homes were primed for action.

Quick reactions. An infected tourist from Wuhan became Hong Kong's first case, but individuals and organisations did not need to wait for official instructions from the government to fight the nascent threat. When the government announced the emergency phase of its infectious disease protocol four days later, nursing homes were already putting their plans into action. Workers had their leave curtailed, preventing them from taking weekend trips to mainland China. Bringing the virus back was less of a possibility in this scenario.

Emergency drills. Not only did nursing homes already have a trained infection controller, but they also carried out emergency drills simulating an infection outbreak four times a year.[14] Infection control was far more likely to become second nature with the regular practice. This 'war footing' encouraged the implementation of a rigid set of measures, reflecting the seriousness of the threat. Infection control policies were based on SARS, rather than influenza, increasing 'battle readiness'.

Taking no chances. Instead of just taking the temperature of all visitors to nursing homes, they banned them altogether – a wise move considering that virus carriers can be asymptomatic. Confirmed cases were also quarantined for up to three months to stop the spread from hospitals into nursing homes. Those who had come into close contact were isolated in a separate quarantine centre for 14 days. And a supercomputer was used to trace the close contacts of infected people to help control cluster outbreaks.

Shoring up the defences. To reduce the risk of infection, nursing home residents were not taken to hospital for medical visits without good reason. This was made possible because after SARS, the Hong Kong authorities increased the capacity of its visiting doctor programme for homes. Also, the effective provision of masks and other protective equipment was embedded into the overall strategy. Hong Kong experts had been advocating the use of face masks for years. Minimal public resistance meant that 70 per cent of Hong Kong residents were already wearing them in January. Most nursing homes had between one to three months' supply of protective equipment.

Mental health. The social restrictions could have impacted very negatively on nursing home residents' mental health. In the absence of social visits and activities, it was recognised that a sense of emotional connection would need maintaining. Healthcare workers were able to arrange WhatsApp video calls so residents could see their family members again. By facilitating the elderly's use of technology to reduce loneliness, their mental health was effectively prioritised.

Making sacrifices. The broader goal of saving the healthcare system from collapse was widely understood. It was therefore essential to ring fence resources for the elderly, since they were more likely to contract the virus, be admitted to hospital, and need a ventilator. The rigid social restrictions needed to be enforced to reduce the risk and spread of infection. Of course, this was by no means easy. The Facebook page for the China Coast Community care home reflected how events had taken their social toll. "We are missing our visits from family, friends, and volunteers," said a post from February.[15]

Judging from the results, it was a price worth paying.

KEY POINTS

- Rewiring our brains for positivity means overcoming the brain's innate negativity bias.
- Speaking up is usually associated with negative events, but it can be truly transformative when the balance shifts in favour of reporting positive ones.
- A focus on positive events creates trust and requires a different kind of mindset. Since the brain is like Velcro for negative experiences, and Teflon for positive ones, we must put conscious effort into focusing on the positive.
- The golden ratio for positive to negative events is six to one.
- A positive script for incident-free safety performance can be created if we first acknowledge the health and safety risks present, and then enact a plan to control them. The royal wedding example provides a blueprint for how this can be achieved.
- Good relationships and close collaboration have a positive impact on safety performance and can pay financial dividends.
- We can learn from the positive experience of others to create a safer experience at all levels, as the Hong Kong nursing homes example shows.

M4: APPLYING MINDFUL SAFETY TO LEARN FROM POSITIVES

Individual

- Our brains are wired for fear. We therefore need to put conscious effort into overcoming the negativity bias and report positive events as much as possible.
- We should aim for six reports of a positive event for every negative one.
- We can all play a part in creating a positive safety script for ourselves and others.
- It is important to remember that there is a huge amount to learn from positive events.

Relational

- Relationships play a critical role in reducing safety incidents. Working together to identify and mitigate health and safety risks can massively reduce the probability of harm or injury.
- Positive, healthy relationships foster an atmosphere of trust for improved reporting.
- Sharing the lessons from positive and negative events can help to increase the level of engagement between managers and frontline operatives.
- Close collaboration to improve the safety environment can positively affect the bottom line.

Organisational

- A special organisational effort to invite the reporting of positive events is needed. It is unlikely to happen without an intervention.
- Reporting forms will need to embody the golden ratio in their design. This will enable the collection of a whole new category of positive data.
- Organisations can maximise the learning from positive events by ensuring they have the forums and channels to communicate important lessons and findings.
- By focusing on the factors that positively maintain high levels of safety performance, organisational learning can help build resilience.

Societal

- Learning from positives is incredibly important at the societal level – this is especially true, for example, in countering the threats posed by a global pandemic such as Covid-19.
- Safety approaches must be prepared to learn from the positive experience of other nations, governments, and societies.
- A transnational perspective is required to deal with common public safety threats of a global nature.
- Shared learning and cooperation across national borders are essential in this context.

NOTES

1. Baumeister, R.F., Finkenauer, C. & Vohs, K.D. (2001). Bad is stronger than good. *Review of General Psychology*, Volume 5(4), p. 333. Available from http://assets.csom.umn.edu/assets/71516.pdf (Accessed 01.06.2018).
2. Hanson, R. (2009). *Buddha's Brain: The Practical Neuroscience of Happiness, Love, and Wisdom*. Oakland, CA: New Harbinger Publications.
3. Gottman, J. (1995). *Why Marriages Succeed or Fail. And How You Can Make Yours Last*. New York: Simon and Schuster.
4. Zenger, J. & Folkman, J. (2013). The ideal praise-to-criticism ratio. *Harvard Business Review*. 15 March 2013. Available from https://hbr.org/2013/03/the-ideal-praise-to-criticism (Accessed 11.08.2018).
5. Federal Railroad Administration (2015). *Continued Improvements at One C^{3}RS Site. US Department of Transportation*. Research Results, pp. 15–17.
6. Ibid., pp. 15–17.
7. Kelly, H. (2018). Royal wedding 2018 viewing figures: how many people watched Meghan Markle marry Harry? *Express*. 20 May 2018. Available from https://www.express.co.uk/showbiz/tv-radio/962610/Royal-Wedding-viewing-figures-Meghan-Markle-Prince-Harry-kiss-David-Beckham (Accessed 02.06.2018).
8. O'Brien, Z. (2018). Prince Harry and Meghan Markle's big day could cost taxpayers more than £30 million because we're picking up the tab for security. *The Daily Mail*. 17 May 2018. Available from http://www.dailymail.co.uk/news/article-5741973/Prince-Harry-Meghan-Markles-big-day-cost-taxpayers-30million-security-costs.html#ixzz5HIAUr0Vf (Accessed 02.06.2018).

9. Royal wedding security is a 'Massive Headache'. *Esquire*. 19 May 2018. Available from https://www.esquire.com/news-politics/a20752256/royal-wedding-security/ (Accessed 02.06.2018).
10. Giannangeli, M. (2018). Royal wedding security: how SAS is preparing for Meghan Markle and Prince Harry's big day. *Express*. 28 April 2018. Available from https://www.express.co.uk/news/royal/952587/royal-wedding-meghan-markle-prince-harry-sas-security-royal-wedding-news (Accessed 10.06.2018).
11. Davidson, T. (2018). Rooftop snipers protect Prince Harry and Meghan at royal wedding as police form ring of steel around Windsor. *Mirror*. 19 May 2018. Available from https://www.mirror.co.uk/news/uk-news/rooftop-snipers-protect-royal-wedding-12560418 (Accessed 02.06.2018).
12. Twomey, J. (2018). Royal Wedding security: police on alert for Meghan and Harry fanatics and stalkers. *Express*. 17 May 2018. Available from https://www.express.co.uk/ news/royal/961011/security-royal-wedding-prince-harry-meghan-markle-stalker (Accessed 07.06.2018).
13. Rushe, D. (2020). Fauci: coronavirus pandemic that 'took over the planet' is far from over. *The Guardian*. 9 June 2020. Available from https://www.theguardian.com/world/2020/jun/09/anthony-fauci-coronavirus-pandemic-far-from-over (Accessed 12.06.2020)
14. Booth, R. (2020). MPs hear why Hong Kong had no Covid-19 care home deaths. *The Guardian*. 19 May 2020. Available from https://www.theguardian.com/world/2020/may/19/mps-hear-why-hong-kong-had-no-covid-19-care-home-deaths (Accessed 12.06.2020).
15. Chor, L. (2020). How Hong Kong avoided a single coronavirus death in care homes. *The Independent*. 26 May 2020. Available from https://www.independent.co.uk/news/world/asia/hong-kong-coronavirus-care-home-death-toll-china-wuhan-covid-19-a9532506.html (Accessed 12.06.2020).

11 From Blame to Safety Enlightenment

It is a capital mistake to theorise before one has data. Insensibly one begins to twist facts to suit theories, instead of theories to suit facts.

– Sherlock Holmes[1]

You may recall from an earlier chapter that blame was the top reason for people not reporting in a healthcare environment. No wonder! Reflect back to a time when you were blamed for something, whether you were responsible for it or not. It might have been stealing a biscuit as a kid, forgetting to do your homework, or as an adult missing a work deadline. I very much doubt it felt good. And I very much doubt it produced a change for the better, although it may have temporarily stopped you from doing something perceived as 'bad' by others.

Blame has unintended negative consequences – it usually makes people clam up, or even shut down psychologically. They may feel ashamed or as if they are enveloped in an atmosphere of fear. This chapter focuses on how to obliterate it, shifting to a fundamentally different mindset in order to create healthier relationships and safer working environments. If you want to prevent accidents, blaming individuals will contribute very little to a more mindful safety culture with low injury rates.

Thinking at the societal level, blame also helps to keep us hooked on the 24/7 news cycle. This is where we turn to next, with the classic example of someone widely blamed in the media for an accident. News organisations often display a malicious tendency to find an individual and flimsily draw a target around them with an incredibly low standard of evidence. We will therefore need the diligence, insight and perseverance of a Sherlock Holmes to unpack the whole subject matter.

Case Study: The Curious Incident of the Train in the Night

The lengths people will sometimes go to blame others is truly astonishing. And of all the places that blame flourishes, it is the media that naturally comes to mind as that crucible for naming and shaming.

The kind of sensational scapegoating I am talking about can effectively bury the truth, at least until people sober up to more considered interpretations of why an accident may have happened. The true facts always take time to emerge. Post-accident, in the initial vacuum of speculation, the media's focus will often be on the people involved in the incident – the individuals who happened to be physically present and directly involved. In this furore, sensationalism will trump common-sense analysis more often than not.

Stealing a Train

Take the case of the cleaner who 'stole' a train. On 15 January 2013, the headlines across the Atlantic read, 'Cleaner Steals Train and Hits House in Sweden'.[2] If you thought the British media would adhere to higher standards of journalism with more balanced, objective coverage, you'd be mistaken – the broadsheet and tabloid press failed to evaluate the facts critically. Even the normally fair-minded, temperate Swedes couldn't resist arresting the seriously injured woman, who apparently crashed the train after taking it from a depot and driving it for a mile.

The suspicion was that the cleaner had intentionally endangered the public, though there was no obvious motive. The train operator initially colluded with those sentiments, effectively feeding the media the craved headlines. Its communications manager informed the Swedish newspaper Aftonbladet that "It was a cleaner who stole the train; somehow she managed to get into the cab and got it moving".[3] The train operator may have apologised later, but the damage had already been done.

At this point, common sense should have prevailed. This was neither an act of terrorism nor an equally improbable 'joy ride' in a train through the suburbs of Stockholm. Seriously, what would motivate a 22-year-old woman with a mop in her hand to go for a joy ride on board a train in the dead of night? With or without the benefit of hindsight, it is a clearly ridiculous notion.

Eliminating the Unlikely

Nevertheless, this was the version of events widely shared in the worldwide press, which was full of blame for a woman who, quite frankly, must have been scared to death. Imagine the sheer terror when she came to the horrible realisation that the train was moving and completely out of control. We do not actually know when, or even if, she became fully aware of what was going on around her, because she couldn't remember the chain of events afterwards. The real story behind all the hyperbole was in danger of being lost, but it is actually far more riveting than the sensationalist headlines. So, in the manner of Sherlock Holmes, pipe smoking in the corner of our mouths, we need to studiously attend to the detail of this case. All the facts presented here have been taken from the official report by the Swedish Accident Investigation Authority.[4]

It had been snowing during the day and the temperature was below freezing. A shunter, on duty at the depot where the cleaner was also working, decided that the vehicles shouldn't be left with the brakes applied, because the blocks could freeze to the wheels during the standstill. That would mean the trains wouldn't run the following morning. It was in fact possible to release the brakes with an authorised procedure that did not engage the driver's controls. This procedure, which involves special equipment to provide a 24 V feed into the driver's cab, was knowingly circumvented by the shunter because the required equipment, though present, was not in working order. It is not clear how widespread this unauthorised practice was prior to the accident.

On the night in question, a loose brake block was used to engage the 'deadman's handle', overriding a key safety device that normally would have prevented the train from moving off. At some point, the train control lever was placed in the 'full power' position. In short, the controls in the driver's cab had been arranged to release the brakes and allow the train to gain tractive power. As she was finishing up, a single button press by the unsuspecting cleaner to close the passenger doors was enough to set the train in motion.

As you already know, cleaners do not as a rule drive trains, or have any training for it. With no idea how to stop the train, her fate was sealed and she would come to rest wherever the runaway train did. The train ploughed through the buffers at the end of the line, ending up in a ground floor apartment and leaving the cleaner incarcerated by the wreckage.

Blame Is Sticky

Holmes might have mused that the attribution of blame early on severely blunted the effectiveness of any real analysis and the following accident investigation. It is extremely hard, if impossible, to erase from the public's consciousness the idea that a single person was to blame, even years on. Unfortunately, blame sticks. The cleaner was caught up in a nightmare beyond most people's imagination and will continue to bear the scars.

In summing up the essential facts on which this case pivots, Holmes may have highlighted the fact that all the information needed to prevent this accident was already known before the crash. But there was certainly no evidence of it being discussed openly. Ideally, the circumvention of an authorised safety procedure should surface long before it endangers lives, but if depot workers do not feel able to talk to managers there is an obvious trust problem. The media grossly simplified and distorted matters and what we need is an explanation for why we appear to have this inbuilt tendency to jump to the conclusion that an individual is to blame

---

THE FUNDAMENTAL ATTRIBUTION ERROR

> Don't let us forget that the causes of human actions are usually immeasurably more complex and varied than our subsequent explanations of them.
>
> **– Fyodor Dostoyevsky**[5]

Social psychologists have long been intrigued by how people arrive at their conclusions about what was responsible for an event. In the absence of the full facts, people essentially make judgements, or what social psychologists call 'attributions', to explain the causes of events.[6] Apportioning blame to someone for an accident is one such attribution, as in the runaway train scenario just described. Research has found that we typically focus our attention more on people as the 'origin' of actions, rather than the situations people find themselves in. This phenomenon has been called the Fundamental Attribution Error.[7]

What superficially appears to be true may hide a more complex story – and we can see this at work in the news all the time. For example, on his first day, the UK's

Home Secretary Sajid Javid was photographed apparently adopting a 'power pose' outside the Home Office.[8] With his feet spaced widely apart, it seemed like a deliberate attempt to show toughness and personal ambition. But sensing an opportunity, photographers had actually asked him to take a step forward. Javid was snapped mid-step, literally walking into a photograph that was not of his own making. But that is not the attribution most observers made – as far as they were concerned, Javid had staged the pose all by himself.

In short, we overplay the individual's role in causing a particular outcome, whilst underestimating the situational context. And because there is a failure to appreciate the importance of situational factors, people are generally held responsible for events, even if they had little hand in causing them. This tendency to distort reality matters a great deal if we are talking about the potential causes of accidents.

Acknowledging Our Bias to Blame

Imagine you have just seen a cyclist riding dangerously through a red traffic light. As an observer, you might say, "What a reckless, thoughtless person!" But what would the cyclist say? They may have assessed the situation rather differently. Supposing they were aware of a car behind driving far too closely. Heavy braking to stop in time might have led to their bike being shunted. At first sight, the cyclist failing to stop may seem like a reckless act, but it may have prevented an accident. Whatever the truth, the interpretation you make as an observer is likely to be different from the cyclist's.

When an organisation has a safety incident, individuals are often blamed, despite the best of intentions. Without an assessment of the full situational context, management or systemic failings are likely to be ignored, only for the same situations to arise repeatedly. For example, take the explosion at BP's Texas City refinery, which claimed 15 lives and injured over 180 people in 2015. It may not come as much of a surprise that a vice president later testified that:

> Our people did not follow their start-up procedures. If they'd followed the start-up procedures, we wouldn't have had this accident.[9]

Of course, focusing exclusively on human failures painted a very one-sided picture. The refinery had in fact been built in 1934 and had been badly maintained for a number of years prior to the accident.[10]

Other, more balanced reports pointed to a problematic safety culture: cost cutting and a failure to invest in plant infrastructure, inadequate training and procedures and poor communications. In high-hazard industries, and indeed elsewhere, there are two games people can play out: the blame game or the enlightenment game (as illustrated in Table 11.1). Both are based on a set of assumptions, often unconsciously held by individuals. These tend to be played out in the context of manager-subordinate relationships, and may be reflective of a wider cultural malaise. Blame cultures in particular are often easy to spot and generate a palpable sense of fear amongst employees.

By challenging the 'blame game' assumptions and developing good practice, we can move towards a much more mature, enlightened culture.

TABLE 11.1
The Two Games of Blame and Enlightenment

The Blame Game	The Enlightenment Game
Promotes the idea that employees making errors are 'bad apples'. If they do not actually intend to cause harm, they are very incompetent.	Promotes the idea that employees making errors are usually conscientious individuals under pressure.
Adopts the view that employees basically 'can't be trusted'.	Adopts the view that employees are trustworthy in most situations.
Disciplines or sacks employees who underperform.	Shares the lessons from good practice to help employees who underperform.
Ignores systemic and cultural factors, such as production or environmental pressures.	Seeks to understand how systemic and cultural factors affect performance.
Demoralises and singles people out for the consequences of their errors.	Empathises and learns from people making errors by stepping into their shoes.
Suggests that the frontline exclusively shoulders responsibility for health and safety.	Shares health and safety responsibility equally amongst managers and the frontline.

TO ERR IS HUMAN

> The new view of human error does not see human error as the cause of failure. It sees human error as the effect, or symptom, of trouble deeper inside a system.
>
> **– Sidney Dekker**[11]

Being human, we all make errors from time to time. Some of these, such as turning up for an appointment on the wrong day, forgetting a friend's birthday, or using a permanent marker on a whiteboard, are irritating to say the least. But they are hardly catastrophic and can be fixed with a little social grace – in the case of the whiteboard scenario, a bit of white spirit will do the trick!

Other errors, such as missing a flight, or putting petrol in a diesel car, can be far costlier. Apart from mixing up the fuel for my car, I've made all of these errors – and I'm sure much of the population would own up to these occasional lapses too. But blaming ourselves isn't going to help much. At an organisational or systems level, a culture of blame can quickly develop, stifling a desire to conscientiously report things which are amiss. In turn, this can prevent any organisational learning from taking place. An overemphasis on human error and a failure to address issues at a systemic level can create the conditions for blame, low levels of employee engagement and even accidents.

To prevent the same errors from happening again, we must therefore be mindful of the consequences of focusing exclusively on human error. If blame creeps in and becomes the norm, errors start to become invisible because they simply go unreported. This might save everyone the discomfort, embarrassment and shame associated with criticism in the short term, but it doesn't bode well for long-term system safety.

Fear Rules Invisibly

If managers chase errors down relentlessly, people start behaving defensively. System failings become vulnerable to being framed superficially in terms of the 'who?' rather than the 'why?' Very few managers would set out to create a blame culture, but it tends to creep in unchecked. Fear can rule with an invisible hand. Ironically, the vast majority of managers would say they favour open dialogue, but fear can prevent their employees from raising safety issues in the first place.

Understanding the reasons for human error requires stepping back to take a broader systems perspective. Human error is usually the last link in a chain of systemic factors – there are always other contributory ones that tend to be downplayed, such as production pressure or environmental conditions. Before reaching any judgement about a human operator's culpability, we must therefore strive to empathically recreate in our mind's eye the complex environment they faced.

MOVING FROM BLAME TO ENLIGHTENMENT

The Fundamental Attribution Error helps us to understand how we show a natural tendency to downplay the situational context. If we are mindful of how the Fundamental Attribution Error works in practice, we can at least counteract it to gain a greater, more objective understanding of the whole picture.

In the aftermath of a safety incident, those directly connected with making an error of some sort often end up feeling like victims. They may be blamed by others, blame themselves, or both. This can create a victim mindset. They may feel helpless, acquiescing to the views of their 'accusers', who may wield more organisational power by virtue of their position. The victim mindset can be compounded if those making critical errors face disciplinary action, or worse, the termination of their employment.

On the other hand, managers or investigators may operate from a persecutor mindset. Challenged to find the root causes of an incident, their analysis may fail to go beyond human error at all, ignoring the extent of systemic failings. They may be part of the management structure that helped to produce the error in the first place, and therefore feel unwilling to view the whole situation in a more balanced, dispassionate light.

I have borrowed some of the thinking from Transactional Analysis, developing an alternative model for more effective use in a health and safety context (see Figure 11.1).[12,13] When tensions run high during an investigation, the stage is set for a drama to unfold. Entrenched organisational power can be used to persecute those responsible at the 'coalface' for making errors. Of course, this is destructive, usually setting the scene for repeat patterns of behaviour in the future.

In recognising the persecutor and victim mindsets, we can take steps to correct them with more enlightened attitudes. There is no need to wait for a safety incident to occur before healthier mindsets are adopted, either. Because of their gravity, and potential for injury and loss of life, safety incidents can magnify the Fundamental Attribution Error, and exacerbate the persecutor and victim mindsets. Seeking out the attitudes responsible for each mindset should ideally happen long before any incident occurs.

The idea here is that the persecutor mindset should be replaced with an influencer one. Instead of attempting to find what is wrong in the behaviour of frontline staff, the overriding principle is that one must strive to see the situation the way they did before the error was made. What pressures were they under? How did they feel? What systemic factors contributed to the error? In this context, error is seen as simply a symptom of deeper issues within the system, not a root cause. If occupying a position of some authority, once they have a clear picture of what needs changing, a manager or safety professional can use their influence to improve the health and safety culture.

The victim mindset needs to be replaced with a participant mindset. This moves away from helplessness and an inherent vulnerability, to a feeling of greater empowerment and a belief in one's ability to co-create a better health and safety culture. In the participant mindset, human error is seen as an inevitable product of the system, not a reason to blame oneself. There is concern rather than guilt about making errors, with a view to assisting managers and safety professionals to reduce them.

This contributes to the creation of a non-punitive culture, where people can speak up honestly about their mistakes without the fear of being reprimanded. The prerequisites for all this to occur are trust and authentic listening.

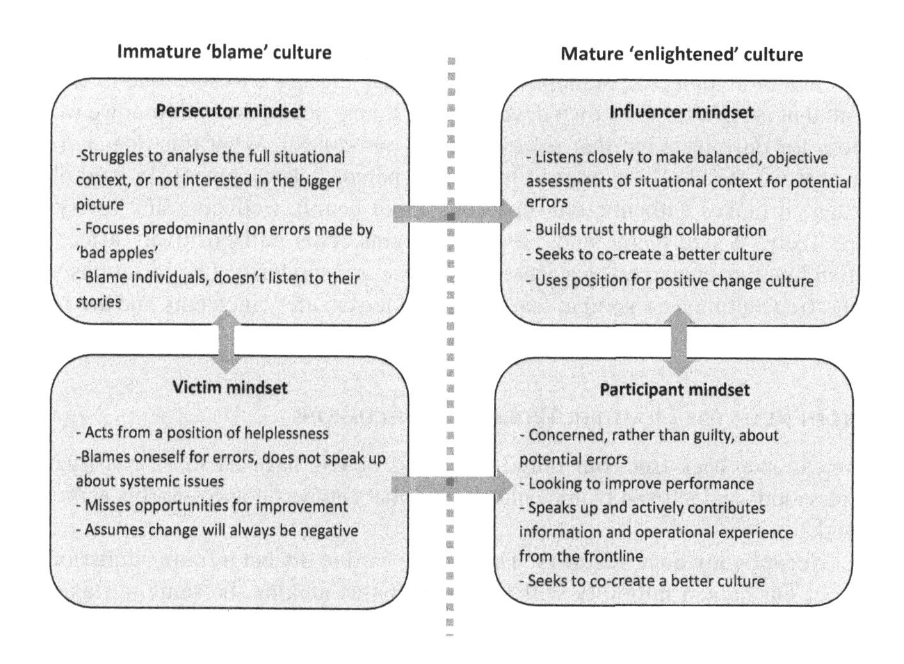

FIGURE 11.1 Shifting mindsets to improve health and safety culture (Langer 2018).

Would You Blame a Lettuce?

> When you plant lettuce, if it does not grow well, you don't blame the lettuce. You look into the reasons it is not doing well. It may need fertilizer, or more water, or less sun. You never blame the lettuce. Yet if we have problems with our friends or our family, we blame the other person. But if we know how to take care of them, they will grow well, like lettuce. Blaming has no positive effect at all, nor does trying to persuade using reason and arguments. That is my experience. No blame, no reasoning, no argument, just understanding. If you understand, and you show that you understand, you can love, and the situation will change.
>
> **– Thích Nhất Hạnh**

Thích Nhất Hạnh's quote is applicable in many situations, at work and at home. It is a reminder to look beyond human error and focus on the bigger picture. In relationships, we typically blame someone for something that has gone wrong. Of course, this is an easy way to channel our anger or frustration, despite our better instincts and the inkling that it will damage the relationship. Often in these cases, some of the underlying desire to blame stems from a belief that someone else must change their behaviour.

In actual fact, we have no control over whether someone else can change. It can therefore be a relief to relinquish control in this respect, because we only have power over one half of the relationship – our half. If we wish to rise above blame, we will need to give up a sense of righteousness or feeling aggrieved in order to better understand another person's perspective. Rather than react in haste when something has gone wrong, we need to detach from the desire to blame.

From a more objective standpoint, we can give the space to someone to choose a path that is right for their own development. It may not be the path that we would choose for ourselves, but that is never under our control. What this does is reaffirm that we are vitally interested in another person's development. In workplace settings, it makes authentic conversations about health, wellbeing and safety far more likely. Where blame ends, learning begins. This will positively affect the bottom line too, encouraging a sense of purpose and solidarity. Organisations with blame-free cultures are good at learning, have fewer safety incidents and are more prosperous.

Golden Rules for Healthier Workplace Discussions

If we can step back from our initial reactions, we can have far more constructive conversations and jettison blame culture. So, what can we do to foster this approach at work?

- **Accept your own mistakes**. This can be hard to do, but a frank admission of our human fallibility will guard against us making the same mistakes again.
- **Accept colleagues' mistakes**. Be compassionate. Were they stressed or tired, and had they had the appropriate training to be able to carry out the task?

- **Avoid defensiveness**. By managing our reactions, we can make our points constructively in the spirit of open dialogue.
- **Learn the lessons**. Shifting the focus to the potential lessons that can be learned from a situation will help ensure a healthier, safer working environment.

KEY POINTS

- In the absence of facts, the media will go to absurd lengths to blame individuals. This must be counteracted with a desire to uncover systemic failures.
- Blame is sticky. In other words, it attaches to individuals who are often vilified for errors they are not responsible for.
- We have an inbuilt, cognitive bias to believe that people cause events, downplaying the situational context. Social psychologists call this the Fundamental Attribution Error.
- There are two games people can play: the blame game and the enlightenment game. It is easy to slip into the blame game as it is often based on unconscious beliefs and assumptions.
- To move from a blame culture to an enlightenment culture, mindsets need to change. Victims must become participants and persecutors must become influencers.
- To effect the change in mindsets, we must learn to accept our own and others' mistakes, whilst focusing on what we can learn from them.

M4: APPLYING MINDFUL SAFETY FOR ENLIGHTENMENT

Individual

- We must remain mindful of our cognitive bias, which distorts reality to view people as the cause of events.
- The Fundamental Attribution Error can be overcome by appreciating the situations individuals find themselves in, not just their actions.
- To help create a mature, enlightened culture in the workplace we can aim to become participants (rather than victims) and influencers (rather than persecutors).

Relational

- In relationships, we need to be mindful of our natural tendency to blame others for things that go wrong.
- In general, we need to pay far more attention to situational factors and empathise with people at the 'sharp end'.
- Learning soft skills, such as authentic listening to promote open dialogue, will help create the right conditions for an enlightened culture.

ORGANISATIONAL

- How an organisation reacts to a safety incident speaks volumes about its approach to learning, and it needs to be monitored closely.
- Does it seek to blame individuals when something goes wrong or does it support them?

SOCIETAL

- An enlightened, blame-free culture can be promoted if managers demonstrate the right attitudes and behaviours.
- A tendency to blame is often promoted by the media and is liable to get inside our heads.
- We can question the basis of the facts we hear, and challenge the narratives presented to us by the media.
- As consumers of the media in various forms, we can resist being swayed by fear and demand more objective coverage to establish the full facts.

NOTES

1. Conan, A.C. (2014). *A Scandal in Bohemia: The Adventures of Sherlock Holmes*, Volume 1. Scotts Valley, CA: CreateSpace.
2. Cleaner steals train and hits house in Sweden. *Sky News*. 15 January 2013. Available from https://news.sky.com/story/cleaner-steals-train-and-hits-house-in-sweden-10457810 (Accessed 16.05.2018).
3. Day, M. (2013). Swedish woman crashes train after taking it for a joyride. *The Daily Telegraph*. 15 January 2013. Available from https://www.telegraph.co.uk/news/worldnews/europe/sweden/9802333/Swedish-woman-crashes-train-after-taking-it-for-a-joyride.html (Accessed 16.06.2018).
4. Swedish Accident Investigation Authority (2014). *Slutrapport RJ 2014:03*. Available from https://www.havkom.se/assets/reports/Swedish/RJ2014_03.pdf (Accessed 18.01.2018).
5. Dostoyevsky, F. (2004). *The Idiot*. London: Penguin Books.
6. Heider, F. (1958). *The Psychology of Interpersonal Relations*. New York, NY: Wiley.
7. Ross, L.D. (1977). The intuitive psychologist and his shortcomings: distortions in the attribution process. In Berkowitz, L. (Ed.), *Advances in Experimental Social Psychology*, Volume 10, pp. 173–220. New York: Academic Press.
8. Crerar, P. (2018). Face of the modern Tory party: Sajid Javid tipped for the top. *The Guardian*. 22 June 2018. Available from https://www.theguardian.com/politics/2018/jun/22/face-of-the-modern-tory-party-sajid-javid-tipped-for-the-top (Accessed 22.06.2018).
9. Calkins, L.B. & Fisk, M.C. (2007). BP official blames Texas blast on workers. *The International Herald Tribune*. 17 September 2007.
10. Reason, J. (2016). *Organizational Accidents Revisited*. Farnham: Ashgate.
11. Dekker, S. (2006). *The Field Guide to Understanding Human Error*. Aldershot: Ashgate.
12. Karpman, S. (1968). Fairy tales and script analysis. *Transactional Analysis Bulletin*, Volume 7(26), pp. 39–43.
13. Berne, E. (1964). *Games People Play*. London: Penguin.

12 Mindfulness Training for Improved Safety Performance

> ...research has demonstrated that mindful people tend to make more accurate judgements, display high problem-solving abilities, and have high task performance.[1]

Everything in this book has emphasised how mindfulness points the way to improved health and safety outcomes. This applies whether we are talking about minimising distractions and concentrating better, working more productively with less conflict, focusing on the right organisational priorities for health and wellbeing, or challenging societal norms. As previously mentioned, compelling brain scan evidence highlights grey matter growth (neuronal cell bodies and synapses) in critical areas after mindfulness training.[2] It works because the brain can be taught to create new neural pathways. In the safety domain, a greater degree of mindfulness is likely to translate into fewer workplace accidents and injuries. Later on in this chapter, we will be looking at mindfulness training interventions that have positively impacted on safety performance. Even during the writing of this book, new research has been further establishing the significant relationship between mindfulness and safety performance. One study of a large petroleum distribution company, which surveyed 706 employees, concluded that mindfulness is an important predictor of safety behaviours.[3] Mindfulness was positively correlated with safety participation and safety compliance, whilst being negatively correlated with workplace injuries.

Unlike a relatively fixed personality trait, mindfulness is far more malleable and can be developed to improve safety performance. This presents exciting new opportunities for safety-focused organisations with a keen eye on results and the bottom line. Much of the failure of conventional training courses stems from their inability to change entrenched habits. To train the mind to perform at a higher level, self-awareness and personal experience must be given leading roles. This creates the groundwork for both attitude and habit change, essentially 'unfreezing' the mind.

It is here that effective mindfulness training can make a huge difference, helping to release people from unsafe habits and behaviours often performed on autopilot. With the right psychological tools to tackle habits, any changes and new ways of working tend to be more permanent – in turn, this leads to higher performance and safer workplaces.

ARROWS TRAINING

ARROWS training is mindfulness based and has been specifically designed to enhance performance and create safer, healthier workplaces. Uniquely, it takes the key principles and practices from mindfulness and tailors them for safety environments, drawing on a wealth of research over the last few decades. The training manual, exercises, practices, and recordings provide a standardised way of increasing safety performance in six key areas known to be integral to mindfulness practice. Based on the 'gold standard' of Mindfulness-Based Stress Reduction (MBSR) training, it has been developed in close partnership with safety-focused transport, logistics and construction companies.

Attention and Concentration

Attention has been recognised as the key component of mindfulness in the research.[4] The mind can be trained to better focus its attention and improve concentration levels in safety environments. By the same token, we can learn to notice when our minds becomes distracted and take remedial action sooner. These skills underpin safe behaviours and help prevent incidents.

Risk Awareness

People can be trained to be more alert to the health and safety risks in their surroundings. Paying attention to what is happening in the moment involves processing internal stimuli (such as thoughts and perceptions), and the external stimuli present in operational environments.[5] Learning to process both kinds of stimuli more mindfully increases risk awareness, whilst enabling people to control their risky or unsafe behavior.[6]

Resilience

It is possible to train resilience to enable us to 'bounce back' quicker in situations that pose a threat to operational safety or our own wellbeing. High performers from the fields of sports, business, and the performing arts, can all provide positive examples of resilience under pressure. Resilience requires effective emotional self-regulation, which is widely reported to be enhanced by the practice of mindfulness.[7]

Outlook

People often report attaining a new psychological perspective on their work and life through greater mindfulness. They can achieve this by applying new skills to establish a more objective relationship with their thoughts and feelings. In addition, it is possible to learn how to flexibly reframe 'negative' situations in the workplace and beyond. New perspectives and insights emerge more readily since people in a mindful state can be "open to several ways of viewing the situation".[8]

Wellbeing

By teaching people effective coping strategies, it is possible to avoid mental ill health and reduce unwelcome costs. Performance and mental health are intricately linked.

People in a positive state of mind are far more likely to perform well on the job. Key elements from Mindfulness-Based Cognitive Therapy (MBCT) have been built into the ARROWS training programme to help trainees take positive steps if they notice their mood is low. There is a well-researched, significant positive relationship between mindfulness and wellbeing.[9]

Stress and Conflict Reduction

People skills come to the fore here. Reducing stress and conflict promotes effective decision-making, and translates into lower rates of absenteeism, higher engagement levels, and greater employee loyalty. The taught skills of mindful communication keep the dignity of all parties intact. This is especially important in safety critical environments where communication breakdowns can cause accidents.

FOCUSING THE MIND

It can feel like a struggle to focus our minds sometimes. One Harvard University study estimated that our minds wander for around half our waking lives, and even on demanding tasks our mind is not present at least 30 per cent of the time.[10] Whilst this may be fairly inconsequential in many everyday situations, it still represents lost productivity of one kind or another. Think about all the time we fritter away when our attention drifts – we do not listen in a conversation and miss important information, we miss our turning in the car, or we forget to send an email attachment.

Try this simple exercise. Focus your attention on the second hand of a clock face nearby, or a watch on your person. Wait until the hand reaches the 12 on the face, then see if you can follow it all the way around for 60 seconds. Concentrating only on the hand, until it returns to its original position. Were you able to concentrate the whole time with no lapses in concentration? If not, how many times did your mind wander? What took your attention away from the task?

In the ARROWS programme, attention and concentration are covered before the other main topics. In fact, attention and concentration thread their way through the whole course because mindfulness depends on them. With good reason, they are considered prized mental resources, especially in the safety domain.

Of course, the situation is far more serious if a safety critical worker is not fully directing their attention to the task at hand. For example, their mind could wander if they are driving a train, piloting an aircraft or running a nuclear facility. One such example from the rail industry with its associated costs is given below.

The Price of Inattention in the Rail Industry

If a train driver has a serious safety incident like a Signal Passed at Danger (SPAD), there are associated costs of around £29,000 in the United Kingdom.[11] Typically, this includes incident review and driver instruction costs. This can escalate to an astronomical £150,000 in the case of a driver dismissal where a replacement driver must be found. There is also a human cost as SPADs can end a train driver's career in an utterly soul-destroying fashion, which is the worst of outcomes for both the employee and their employer. Society may subsequently end up picking up the additional costs of unemployment and mental ill health, but this is all completely preventable.

Distraction has been highlighted as the most common, immediate cause of a SPAD. Other factors such as fatigue and familiarity (responding to an expectation rather than reading the signal) also play significant roles. The good news is that the effects of distraction, fatigue and familiarity can all be proactively addressed with mindfulness training. It is regrettable that experience suggests that we often wait for an incident before learning the lessons – it needn't be that way if we can fundamentally change our mindset. This book has suggested throughout that by training the mind to focus its attention effectively, conventional, 'focus-on-what-goes-wrong-thinking' can be reversed.

In the SPAD example, it is easy to see how the shift in thinking could impact workplace relationships. The dialogue between a driver and manager can be reframed to support higher levels of concentration. Conversations about sustaining situational awareness take on new meaning, with fear and blame permanently relegated to the psychological dustbin. In an enlightened organisational culture, we will be focusing instead on ways to improve attention and performance – and providing the step-by-step means to get there.

This naturally implies a shift in organisational priorities, away from dealing largely with the consequences of incidents to focusing more on how they can be prevented in the first place. If operational employees are taught how to concentrate when incidents do occur, the conversation can turn to how best to improve attention on the job. This is preferable to automatically going down a disciplinary route, which usually raises stress levels for all those involved while achieving little.

Safety Training Applications in Industry

Adopted more extensively by high-hazard industries and beyond, mindfulness has the power to transform training and development for safety critical roles. Whether we are talking about train drivers, pilots, doctors, construction workers, crane operators, oil rig workers or their managers, it can reliably enhance concentration and situational awareness. Its full potential as a training intervention is most likely to be achieved where a strong, mindful safety culture is already present.

There are now significant results from the safety domain which demonstrate mindfulness has the clear potential to reduce risky behaviour and save lives. In one training intervention, London bus drivers reported a statistically significant decrease in risky behaviours.[12] Some examples from this field study are presented below.

Case Study: Mindfulness on London's Buses

An accident involving a bus on the streets of London might be far less newsworthy than one involving an aeroplane, but it still has the potential to seriously injure and kill people. A moment's distraction at the wheel could prove fatal for passengers or pedestrians. Drivers often face extremely challenging situations and are also expected to provide a high level of customer service.

The training programme described here was mindfulness-based, but tailored to the specific needs of bus drivers. All 23 drivers who participated in the research were encouraged to listen to mindfulness recordings and practise daily. Through persistent practise, the course aimed to cultivate acceptance, calmness, resilience and empathy. The course embedded new habits over eight weeks and covered the following areas: driving more safety and coming off autopilot, minimising distractions, risk awareness, on-the-job resilience, assertive communication, and stress management.

The results clearly demonstrated the incredibly positive effect mindfulness can have on individuals and the business of driving a London bus:

- 100 per cent said they drove more safely than before the course.
- 95 per cent said they responded to stressful situations better.
- 95 per cent said they gave better customer service.

Here are some of the benefits the drivers reported themselves:

Improved Alertness

- Paying better attention to the road. One driver said: "I've noticed that when I go on autopilot, I can pull myself back out of it."
- Scanning the bus stop on the approach for any hazards or dangerous objects.
- Being alert to the time passengers need to get off the bus, particularly if they are elderly or vulnerable.

Calmer Driving

- Several drivers reported no longer being bothered by other road users cutting them up. They were able to 'let go'.
- One driver reported that he was now driving in a calmer, safer manner. A passenger commented: "I really like your driving!"
- Another driver reported he was using the horn out of anger and frustration far less than he had prior to starting the course.

Calmer Responses to Challenging Situations

- A police car drove dangerously in front of a bus and the driver had to brake hard to avoid an accident. He was shaken up and pulled over angry and upset. Using the three-minute breathing exercise to calm his mind, he was able to return to driving. "I would never have been able to do that before the course," he said. The police car driver returned to the scene to offer an apology, admitting he had lost his concentration.
- One driver was insulted by a passenger, but was able to use the three-minute breathing exercise to de-stress. She was then able to continue her driving in a calmer state, feeling re-energised.
- A passenger began screaming at the driver, accusing him of missing her stop. The passenger had actually made a mistake, as it was the stop for another route. The driver was able to explain once the passenger had calmed

down, leading to her offering an apology. He said he would have previously reacted angrily and had been prone to getting into arguments. Increased self-awareness gave him new options to diffuse such situations.

Beyond the Workplace

- There are recognised health benefits to doing mindfulness-based safety training. One driver said he had stopped taking strong painkillers for severe migraines. He has found the body scan more effective at relieving his pain and there are none of the side effects.
- One driver commented that his family had been pleasantly surprised by his calmer driving. He was no longer rushing around.
- Another said the course was giving her the skills to effectively manage stress levels, such as when caught up in traffic on the way to work.

Considering the costs of becoming distracted at the wheel of a bus, or the controls of a train, which can easily run into tens of thousands, the cost of training people to improve their attention and concentration represents a very modest outlay.

Mindfulness-based programmes have had hugely positive impacts in other safety-focused environments too. In one healthcare setting, 61 nurses trained in mindfulness demonstrated a significant reduction in job burnout, as well as an increase in self-compassion and serenity.[13] Though the benefits reported here were mainly related to staff well-being, research on the positive contribution of mindfulness to work performance is beginning to accumulate.[14] This of course makes intuitive sense, because when we feel less stressed and happier in ourselves we are likely to perform better too.

In more complex safety-focused environments, mindfulness programmes can deliver not just improvements in wellbeing, but in attention-related task performance too. One such study carried out at a UK nuclear facility shows how this can be achieved in practice. Statistically significant results were achieved for attention-related and wellbeing measures.[15]

Case Study: Mindfulness in Nuclear Power

With a clear focus on two pillars of mindfulness practice – attention and wellbeing – staff at a nuclear facility were trained in mindfulness over an eight-week period. In total, 100 staff were trained in different groups over a two-year period. The programme was tailored for staff from a range of departments and grades within the host organisation to support the overall aim of improving human performance. A control group was used to increase the robustness and validity of the study.

One of the challenges was designing a course that would appeal to a largely male technical audience. As one participant put it:

> Initially, I was apprehensive and even quite dismissive of mindfulness, but I'd heard good feedback from those who had gone to the previous courses so thought I would try

> it. How wrong I was. I found it really insightful, and am now a huge advocate. I have continued the home practices as they are so beneficial, and I'd highly recommend the course to everyone.

Staff attended the programme voluntarily, participated in group sessions to encourage a high level of engagement, and undertook daily home practice. A range of personal benefits were reported by participants, including increased levels of general wellbeing and resilience. More specific examples, such as being less self-critical, as well as the sense of having a greater range of choices in response to situations, were frequently reported too.

Three areas of workplace benefit came to the fore in personal evaluations of progress:

- Being more present and less distracted, more focused and more able to concentrate.
- Being more able to stand back and prioritise.
- Supporting others through increased levels of empathy as well as patience, and engaging more fully with colleagues.

Self-reported wellbeing levels rose by 22 per cent amongst participants, whilst attention levels were 34 per cent higher at the end of the course (both results were at statistically significant levels).

In their own words, here are some of the comments made by participants on the course, organised by the categories: attention, stress and conflict reduction, and outlook.

Attention

- "In terms of the workplace I find that I am more focused on what I am doing. I can prioritise better and attend to one job at a time rather than getting distracted by other jobs part way through."
- "I've taken steps to eliminate distractions and focus on my intended task. I've also been more able to cope with prioritization and dealing with not being able to please everyone."
- "I am able to employ some of the breathing and attentive techniques in order to break out of a negative spiral. I find that I'm more able to concentrate upon individual tasks at work."

Stress and Conflict Reduction

- "As a leader at work the training has given me excellent insight to help my staff, and also for me to behave in a way which will be supportive."
- "I have found that I am able to deal better with clashes in personalities: largely by being able to accept or acknowledge differences and move on. I have found empathy has been enhanced. I have become more sensitive to my staff's stress levels or struggles and seeing the signs."

- "I have learned to be more forgiving of myself and to understand that there is always a reason why other people act the way they do, and that I should be more forgiving of them."

Outlook

- "The course has given me a whole new perspective on life. Just a few simple concepts and a bit of discipline on my part has really improved my quality of life."
- "The course has helped me realize that it is possible to take a step back from day to day worries and make time for myself without feeling guilty."
- "Being able to be curious about a situation has allowed me to choose to react and ultimately feel differently about certain situations."
- "Attending these sessions has turned my life on its head and I can honestly say that it's made me much happier, more focused and much more resilient."

Beyond the Workplace

- "I feel happier and more able to achieve what I really desire in work and my personal life."
- "I have noticed a significant improvement in emotional awareness, personal wellbeing and overall happiness. One particular bonus is that I have become much more patient! All of the above have had an impact in the workplace."
- "Attending these sessions has turned my life on its head and I can honestly say that it's made me much happier, more focused and much more resilient."

TRAINING FOR PREVENTION

I once saw some video footage of commercial drivers microsleeping and being distracted at the wheel. Of course, the video evidence obtained from cameras in the cab was indisputable. It confirmed those drivers did not have their attention on the road. But why wait until events like this show up at the sharp end? The emphasis on catching those drivers 'red-handed' in the middle of cognitive lapses suggested driver wellbeing did not feature highly on the safety agenda.

Prioritising health and wellbeing before it gets to that stage will likely translate into fewer incidents, happier employees and a much stronger bottom line. A preventative mindset is needed – this is where organisational mindfulness has a strong role to play. Had those 'offending' drivers been trained effectively, they might have displayed a much higher degree of alertness, decreasing the chances of an incident. This is all about reframing organisational objectives to systematically focus on what is going right, rather than looking almost exclusively at what is going wrong.

Our questions about human performance need reformulating to reflect a desire to learn from positives. For example, what can we learn from employees who have achieved high-quality sleep patterns and can effectively manage their fatigue? How can the knowledge of staying alert that some drivers evidently possess benefit their colleagues who find it more difficult? Questions like these can be asked well before there is a safety incident, creating the groundwork for far more positive safety outcomes.

Mindfulness-based training may give businesses an advantage in helping to answer some of these questions. Google's example shows how mindfulness can gain traction in a high-tech corporate environment. By appointing a head of mindfulness training, the business recognised some time ago that success is dependent on happy workers with healthy, alert minds.[16] The same thinking, albeit with a sharper emphasis on safety, is now being applied in high-hazard industries. The trend can only grow as safety-focused organisations increasingly value mindfulness-based safety training for the benefits it brings – not just in terms of enhanced wellbeing, but in terms of improved attention and concentration levels.

KEY POINTS

- Evidence from the neuroscience shows how the brain changes physically as a result of daily mindfulness practice.
- Over the last few decades, mindfulness programmes have consistently demonstrated their effectiveness in improving the wellbeing of participants, but there is now a strong interest in improving safety outcomes too.
- The ARROWS programme takes the essential elements from traditional mindfulness programmes, but specifically tailors the content for safety environments.
- The business case for mindfulness training in frontline roles is far easier to make when the costs of safety incidents caused by inattention are acknowledged. The training costs are often small in comparison.
- Where ARROWS style mindfulness programmes have been introduced in safety environments, the results have proven to be statistically significant in the areas of attention-related performance and the reduction of risky behaviours.
- Research from real-world environments (e.g. London's buses, nuclear power and healthcare) shows how mindfulness programmes can positively impact safety in a variety of settings.

M4: APPLYING MINDFULNESS TO IMPROVE SAFETY PERFORMANCE

INDIVIDUAL

- Individuals who practice mindfulness report higher levels of wellbeing, and higher levels of attention and concentration.
- In safety environments, this translates into higher performance on safety-related tasks.
- Learning how to embed new habits (and give up counterproductive ones) helps create safer behaviours.
- Many participants on mindfulness programmes also report a fundamental shift in life perspective.

Relational

- In the calmer state of mind taught by mindfulness, relationships become easier to manage with less interpersonal tension.
- Mindful people tend to be more empathic, increasing the chances of successful conflict resolution.
- More options for resolving differences become apparent, since mindfulness reduces defensiveness or reliance on strongly held views.
- Mindfulness encourages assertive communication, helping to replace passive or aggressive styles.

Organisational

- Mindfulness training programmes tailored for safety environments, such as ARROWS, can play a key role in reducing the likelihood of a safety incident at work.
- The research evidence also suggests such programmes can bring about positive change in terms of safety participation and safety compliance.
- These programmes adopt a preventative mindset, an important factor in creating a more mindful safety culture.
- Their appeal lies just as much in improving attention and concentration, as it does in enhancing levels of wellbeing.

Societal

- The economic and social costs of safety incidents to society can be reduced by the greater adoption of mindfulness training.
- It also encourages safety awareness outside work, which is especially important in the current climate, where safety is a responsibility shared by whole populations.
- More mindful employees are likely to make more mindful citizens, valuing not just their contributions to work, but to wider society too.
- Mindfulness fosters gratitude and an appreciation for the interconnectedness of our lives at a societal level.

NOTES

1. Kao, K., Thomas, C.L., Spitzmueller, C. & Huang, Y. (2019). Being present in enhancing safety: examining the effects of workplace mindfulness, safety behaviors, and safety climate on safety outcomes. *Journal of Business and Psychology*.
2. Hölzel, B.R., Carmody, J., Vangel, M., Congleton, C., Yerramsetti, S.M., Gard, T. & Lazar, S.W. (2010). Mindfulness leads to increases in regional brain gray matter density. *Psychiatry Research: Neuroimaging*. doi:10.1016/j.pscychresns.2010.08.006. Epub 2010 November 10.
3. See note 3 above.

4. Brown, K.W. & Ryan, R.M. (2003). The benefits of being present: mindfulness and its role in psychological well-being. *Journal of Personality and Social Psychology*, Volume 84(4), pp. 822–848.
5. Brown, K.W., Ryan, R.M. & Creswell, J.D. (2007). Mindfulness: theoretical foundations and evidence for its salutary effects. *Psychological Inquiry*, Volume 18(4), pp. 211–237.
6. Feldman, G., Greeson, J., Renna, M. & Robbins-Monteith, K. (2011). Mindfulness predicts less texting while driving among young adults: examining attention and emotion-regulation motives as potential mediators. *Personality and Individual Differences*, Volume 51(7), pp. 856–861.
7. Brown, K.W., Ryan, R.M. & Creswell, J.D. (2007). Mindfulness: theoretical foundations and evidence for its salutary effects. *Psychological Inquiry*, Volume 18(4), pp. 211–237.
8. Brown, J. & Langer, E. (1990). Mindfulness and intelligence: a comparison. *Educational Psychologist*, Volume 25(3–4), pp. 305–335.
9. Pidgeon, A.M. & Keye, M. (2014). Relationship between resilience, mindfulness, and psychological well-being in university students. *International Journal of Liberal Arts and Social Science*, Volume 2(5), pp. 27–32.
10. People spend 'half their waking hours daydreaming'. *BBC News*. 12 November 2010. Available from https://www.bbc.com/news/health-11741350 (Accessed 19.07.2018).
11. Simanjuntak, S. (2014). *The Indirect Costs Assessment of Railway Incidents and Their Relationships to Human Errors. The Case of Signals Passed at Danger.* Available from https://www.imperial.ac.uk/media/imperial-college/faculty-of-engineering/civil/public/ug/ug-final-year-projects/T13---Samuel-Simanjuntak.pdf (Accessed 16.07.2018).
12. Langer, C. & Plant, K. (2020). Using mindfulness to reduce the risk of a safety incident on London's buses. Unpublished research paper.
13. Penque, S. (2019). Mindfulness to promote nurses' well-being. *Nursing Management*, Volume 50(5), pp. 38–44.
14. Dane, E. (2011). Paying attention to mindfulness and its effects on task performance in the workplace. *Journal of Management*, Volume 37(4), pp. 997–1018.
15. Summerfield, M. (2020). Bringing mindfulness into safety critical environments. Unpublished case study.
16. Confino, J. (2014). Google's head of mindfulness: 'Goodness is good for business'. *The Guardian*. 14 May 2014. Available from https://www.theguardian.com/sustainable-business/google-meditation-mindfulness-technology (Accessed 16.07.2018).

Conclusion

The new approach espoused throughout this book has been all about implementing safety more mindfully at four distinct levels: the individual, relational, organisational and societal. A comprehensive approach to safety has never been more necessary. Our daily lives have been turned upside down down by Covid-19. To fight a pandemic and remain resilient, we have no choice but to adopt new thinking, habits and practices for the benefit of our personal health, wellbeing and safety.

The M4 approach isn't an academic, analytical approach confined to a classroom or a flip chart. Nor is it merely a 'talking safety' approach, though dialogue is, of course, an important ingredient. It is a method for empowering people, a comprehensive set of tools for breaking old habits and enhancing the mind's power to focus. It has a direct application to high-hazard industries where attentiveness to safety critical tasks is essential, but is equally applicable in wider society too. The M4 approach is holistic in both nature and intent, with a multi-level focus on shifting awareness, attitudes, thinking and behaviour.

As an action-based approach, it delivers concrete results to improve the bottom line – fewer safety incidents, better working relationships and greater business focus. Mindfulness-based training programmes tailored for safety environments have produced statistically significant results in healthcare, transportation and energy settings, as the wide range of examples in Chapter 12 demonstrate. There are few other training interventions backed by science that have such a positive impact across a range of measures – from reducing stress levels and enhancing wellbeing, to improving attention levels, and broadening one's outlook on life well beyond the workplace.

Embracing a mindful safety culture means collectively focusing more on cognitive successes than failures. It means expanding our thinking beyond the limited analytical power of human error or incident classification systems. Once mindfulness becomes a habit, people start to notice the positive difference it makes in their everyday lives. It is the experience that counts, often encapsulated in that moment when someone says: "I can see the change within me, so I *know* it works." 'Knowing' in this sense means far more than understanding something intellectually. This process of operationalising knowledge is greatly enhanced at a social level, where dialogue is both possible and encouraged.

I have often been taken aback by the results obtained in group settings. The effect of safety professionals sharing experiences with each other tends to reaffirm their expertise, whilst reinforcing the belief that they can make the safest possible decisions in real time. In an atmosphere of trust, critical incidents and errors can be discussed without blame or censorship, and with a genuine desire to learn from them. But, as has been frequently emphasised, it is learning from positives that represents the greatest learning opportunity of all.

Freed from the traditional constraints of bureaucracy, participants on mindfulness programmes are able to draw on their own resources. The skills they learn can give them more options to respond to the situations they face – at work and at home – and encourage more flexible thinking. Focusing the mind is just as important for bus and train drivers, as it is for pilots, doctors, nurses, miners or construction workers. The flexibility of recognised mindfulness programmes, such as ARROWS, to accommodate a wide range of safety settings, and grades of staff, is an important attribute.

Successful training interventions will tend to become part of organisational culture, implemented without the need for an overdose on corporate soundbites or hubris. They involve changing how people talk, think and behave, but above all, what they do *after* the training has been completed and operationalised. Uniquely perhaps, mindfulness teaches how unhelpful, old habits can be defeated and new ones formed. Most training courses fail in this respect, as their content is forgotten soon after attendance. In contrast, short-term memories are consolidated into long-terms ones through the daily practice of mindfulness.

Speaking more broadly, M4 approach is a root and branch approach with the goal of changing how we practice health, wellbeing and safety. As such, it is inseparable from organisational culture and business performance. Because the approach focuses its attention evenly between four levels, it is neither 'top-down' nor 'bottom-up' in essence. It is freed from the constraints of this false, hackneyed dichotomy from the safety thinking of yesteryear. It adopts a preventative mindset, and fosters greater awareness amongst all grades of staff, ultimately being good for business too. In practice, if well implemented, it means fewer safety incidents, reduced absenteeism, greater staff loyalty and lower staff turnover.

There is a vast reservoir of untapped expertise and learning already present in the minds of employees. The task of leadership is to harness, rather than suppress, the energy of the workforce in the pursuit of optimal performance. If we merely strive for employees to arrive at home safe and healthy after a good day's work, we will be missing an opportunity by setting the bar too low.

With the right vision, we can finish each day in the knowledge that our safety, health and wellbeing have tangibly improved because of the work we have done. I hope you will join me in making that vision come to life.

Index

Printed in the United States
by Baker & Taylor Publisher Services